This book is to be returned on or before
the last date stamped below.

The University of Melbourne–
Nucleus Multi-Electrode Cochlear Implant

Advances in
Oto-Rhino-Laryngology

Vol. 38

Series Editor
C.R. Pfaltz, Basel

Basel · München · Paris · London · New York · New Delhi · Singapore · Tokyo · Sydney

The University of Melbourne – Nucleus Multi-Electrode Cochlear Implant

G.M. Clark in collaboration with
P.J. Blamey, A.M. Brown, P.A. Gusby, R.C. Dowell, B.K.-H. Franz, B.C. Pyman, R.K. Shepherd, Y.C. Tong and *R.L. Webb*
University of Melbourne, Melbourne, and
M.S. Hirshorn, J. Kuzma, D.J. Mecklenburg, D.K. Money, J.F. Patrick and *P.M. Seligman*
Cochlear Pty Limited, Sydney

100 figures and 42 tables, 1987

Basel · München · Paris · London · New York · New Delhi · Singapore · Tokyo · Sydney

Advances in Oto-Rhino-Laryngology

Library of Congress Cataloging-in-Publication Data
    Clark, G.M., 1935–
    The University of Melbourne – nucleus multi-electrode cochlear implant.
    (Advances in oto-rhino-laryngology; vol. 38)
    Bibliography: p.
    Includes index.
    1. Cochlear implants. I. Title. II. Series.
    [DNLM: 1. Cochlear Implant. 2. Deafness-surgery. 3. Electrodes, Implanted.
    W1 AD 701 v. 38 / WV 274 C593u]
    RF16.A38 vol. 38 [RF305] 617.8'82 87-16909
    ISBN 3-8055-4575-4

Drug Dosage
    The authors and the publisher have exerted every effort to ensure that drug selection and dosage set forth
    in this text are in accord with current recommendations and practice at the time of publication. However,
    in view of ongoing research, changes in government regulations, and the constant flow of information
    relating to drug therapy and drug reactions, the reader is urged to check the package insert for each drug
    for any change in indications and dosage and for added warnings and precautions. This is particularly
    important when the recommended agent is a new and/or infrequently employed drug.

Contents

Contents

Preface

Cochlear prostheses are one of the first neuroprosthetic devices to be used widely on a regular clinical basis. They have resulted from multidisciplinary research in physiology, biology, surgery, engineering, psychophysics, speech science, and other related fields. There can be few developments that have required such a diverse input. Cochlear prostheses have contributed to the scientific basis underlying the specialty of oto-rhino-laryngology. In the future they should also make major contributions to the practice of audiology, education of the deaf and speech pathology. Although a new era in the management of deafness has emerged there is a great need for continuing research in this field to improve the performance of devices so that as many patients as possible may benefit.

Acknowledgements

The research to develop a multiple-electrode cochlear prosthesis would not have been possible without the financial support of the National Health and Medical Research Council of Australia, the Channel 10(0) Telethon and the Deafness Foundation of Victoria, Lions International and Trusts and Foundations. The industrial development of the prosthesis for clinical trial was funded by a Public Interest Grant from the Australian Government.

We would also like to acknowledge with gratitude the valuable contributions made to this project by our scientific colleagues who include: Dr. R.C. Black, Dr. D.J. Dewhurst, Dr. I.C. Forster, Dr. E. Javel, Ms S. Luscombe, Ms S. Roberts, Prof. G.B. Ryan and Mr. R.J. Laird. Special thanks are due to Mr. John Scrimgeour, Department of Ophthalmology, University of Melbourne, for the photography, Mr. G. Cook, Mr. D. Bloom, Ms Leanne Hurlston and Ms J. Ng for technical support, Mrs. Enid Utton and Mrs. Margaret Gilmour for the typing, and Ms Judy McNaughton for preparing the index.

The artwork for the surgical drawings was done by Mr. Ed Zilberts, and provided by Mr. R. West, president Cochlear Corporation, Englewood Co., USA.

Foreword

The first reports of electric stimulation of the acoustic nerve in man go back to the late 1930s. It took another 20 years until 1957 the French group of Djourno, Eyries and Vallancien were able to publish their results of successful electric stimulation of the cochlear nerve in 2 human subjects. They implanted microinduction coils which stimulated the nerve 'à distance'. However, technical failures discouraged this group of pioneers to continue their research. Seven years later (1964) B. Simmons, Mongeon, Lewis and Huntington from Stanford University Medical School reported their results of direct electric stimulation of the cochlear nerve by means of a multiple bipolar electrode system, implanted in a deaf human subject. Also this research group abandoned their project because the patient's benefit from the cochlear implant was not very promising in the long run. Moreover, experimental data were lacking which indicated that electrodes could be implanted into animal or human cochleae without causing long-term secondary degeneration of the afferent nerve fibers.

In 1968 Michelson was able to give substantial evidence from his animal studies that intracochlear electrodes could be maintained safely and would be functioning for long periods of time. W. House and Urban then began to develop more sophisticated cochlear implant systems. Their extensive studies on a single subject, who had been successfully implanted, guided further engineering development and resulted in the first stimulator package which could be worn by the patient without impeding his professional or social activities. This was the first breakthrough from the merely experimental period into the clinical application period of cochlear implants. In 1973 the House group developed a preoperative diagnostic test battery for selection of patients and a postoperative rehabilitation program. In the same year the first international conference on electric stimulation of the acoustic nerve as a treatment for profound sensorineural deafness in man was held in San Francisco under the presidency of the late F. Sooy, at that time chancellor of the University of California.

All over the world research groups were forming, representing multidisciplinary teams including audiologists, bioengineers, speech and hearing therapists, psychologists, social workers and last but not least otologists, who took the responsibility for selecting, operating and guiding the implanted patients. Technical progress was rather rapid: Not only intra- but also extracochlear implants were developed; single-channel electrodes were replaced by multichannel devices, speech processors became smaller and more efficient. This technical evolution was only feasible by means of progress in electronic engineering on the one side and an intensive neurosensory research on the other.

At the University of Melbourne G. Clark started in 1967 his physiological research on electric stimulation of the cochlear nerve in animals. In 1970 he published a paper on the neurophysiological assessment of the surgical treatment of perceptive deafness, in 1973 he reported the results of his experimental studies in 4 cats: 'A Hearing Prothesis for Severe Perceptive Deafness'. In 1978 a multichannel prototype receiver had been fabricated and the preliminary safety studies completed. The first two implants were carried out in 1978 and 1979 and the results were encouraging. In close cooperation with the Australian biotechnical firm Nucleus Ltd., with a substantial support by the Australian government, Professor G. Clark and his research group from Melbourne University developed a multichannel cochlear implant providing spectral information in addition to temporal and intensity cues, thus allowing profoundly deaf postlingual adults not only to perceive the complex noise of their acoustic environment but also to recognize words with and even without visual cues. The development of this new multichannel cochlear implant may be considered a second breakthrough, closing the period of clinical studies and opening a new area of practical application and clinical testing of cochlear implants. The present monograph provides detailed information about the physiology and psychophysics of electric stimulation of the auditory nerve, about biological aspects (e.g. biocompatibility), the engineering of the receiver-stimulator and speech processor, the selection of patients, surgery, psychophysics and speech perception in postlingually deaf adults. This monograph also gives evidence of the important scientific work accomplished be the multidisciplinary Australian research group which must be considered an important contribution to the rehabilitation of the deaf.

C.R. Pfaltz, Basel

Introduction

A prototype receiver-stimulator and multiple-electrode array developed at the University of Melbourne was first implanted in a postlingually deaf adult patient with a profound-total hearing loss on 1st August 1978 [36]. A speech-processing strategy which could help patients understand running speech, especially when combined with lip-reading, was proposed in 1978 following initial psychophysical results on the first patient [130, 131]. A prototype wearable speech processor was developed in 1979 that could provide significant help to patients in understanding running speech when used in combination with lip-reading compared to lip-reading alone, and also enable them to understand some running speech when using electrical stimulation alone. An implantable receiver-stimulator and wearable speech processor embodying the principles of the prototype device were then produced for clinical trial by the Australian biomedical firm, Nucleus Limited, and its subsidiaries, Cochlear Pty Limited and Cochlear Corporation. This cochlear implant was initially clinically trialled on six patients at The Royal Victorian Eye and Ear Hospital in 1982, and shown to give similar results to those obtained with the prototype device [42, 43]. In view of these results, a clinical trial was carried out for a Pre-market Approval Application to the US Food and Drug Administration (FDA) [146], and extended to a number of centres in the USA, Canada and the FRG. This clinical trial confirmed that patients could understand running speech when electrical stimulation was combined with lip-reading, and that some patients could also understand running speech when using electrical stimulation alone [67]. Today, more than 500 patients world wide are using the cochlear implants developed from the research described in this monograph.

It is now considered appropriate to review some of the work carried out that led to the development of this multiple-electrode intracochlear implant, especially the research that was concerned with physiology, biological safety, engineering, surgery, psychophysics and speech perception. It is beyond the scope of the monograph to adequately review the work of other research scientists as reviews of their work may be obtained elsewhere [102]. Research in the area is also well covered in the proceedings of major conferences on the subject, for example: Cochlear Prostheses: An International Symposium – Annals of the New York Academy of Sciences, volume 405, 1983; Second International Symposium on Cochlear Implants, Paris, Acta Oto-Laryngologica, Supplement 411, 1983; The 10th Anniversary Conference on Cochlear Implants: an International Symposium, University of California, 1983; International Cochlear Implant Symposium and Workshop, Melbourne, 1985, Annals of Otology, Rhinology and Laryngology, Supplement 128, vol. 96, 1987. Cochlear implant research has also been presented in: Cochlear Implants, Ed. Roger F. Gray, Croom Helm, London, 1985, and The Cochlear Implant, ed. Thomas J. Balkany, The Otolaryngologic Clinics of North America, volume 19, number 2, 1986.

The research that was undertaken to develop the University of Melbourne/Nucleus cochlear prosthesis is described under the following sections: physiology, safety, engineering, patient selection, surgery, psychophysics and speech perception. Some of the research in these areas was undertaken in parallel, and some had to be done in series following prior investigations.

Acute physiological studies were first commenced in 1967 to help determine the central neural limitations on rate of electrical stimulation in coding speech information. This work showed that there was an upper limit of about 300 pulses/s for processing electrical stimulation. In 1970 behavioral studies were commenced on the experimental animal to study the perceptual limitations of rate of stimulation, and showed that the animals could not detect changes in rate of stimulation at stimulus rates greater than 600–800 pulses/s. These studies confirmed the limitations of conveying speech by rate of stimulation using a single-channel stimulus alone, and suggested that a cochlear prosthesis should be designed to provide multiple-electrode stimulation. To do this meant implanting a unit in a patient that would contain electronic circuitry that could receive information from outside the body and present the information as patterns of stimulation to the auditory nerve.

The development of a prototype device was commenced in 1971 in conjunction with Dr. David Dewhurst from Bioengineering at the University of Melbourne. Our previous animal experiments had shown the difficulties of using a percutaneous plug, and that infection could result. This was the reason it was considered in patients' interests to develop a unit that was totally implantable. This meant that we would not have complete freedom to determine the appropriate stimulus parameters to code speech information, therefore the unit had to be designed to be as flexible as possible to enable the initial experiments to be carried out.

While the electronic engineering development was being undertaken a series of studies were carried out to determine how best to implant electrodes into the human cochlea, the appropriate design of these electrodes, methods of localizing the electrical current to discrete groups of auditory nerve fibres, and the effects of implanting electrodes in the cochlea on the long-term survival of the spiral ganglion cells. These studies also commenced in 1971.

In 1978 the prototype receiver-stimulator had been fabricated and the preliminary safety studies completed. It was only then that the first patient received his implant, and commenced a series of psychophysical tests to determine the percepts obtained with electrical stimulation. A physiologically based speech processor producing simultaneous stimulation of the auditory nerve was tried in 1978, and found to give poor results. This may have been due to uncontrollable variations in loudness. However, following psychophysical tests, a laboratory-based, feature-extraction, speech-processing system was evolved which was found to provide the patient with significant help in understanding running speech. This was first developed at the end of 1978.

The next task was to validate that this feature-extraction multiple-electrode speech-processing system could help other profound-totally deaf adult patients with a postlingual hearing loss, and this was done on a second patient in 1979.

Following the encouraging results on the first two patients the next challenge was to engineer a speech processor that was small enough to be worn by the patient so that it could be used in their everyday environment. This was completed towards the end of 1979.

Having shown that it was feasible to develop a cochlear implant that could be used by profoundly-totally deaf adults with a postlingual hearing loss a public interest grant was awarded by the Government to the Australian biomedical firm Nucleus Limited to produce, in cooperation with the University of Melbourne, a cochlear implant and speech processor for clinical trial. In addition, the University of Melbourne undertook further research to determine the safety of the device. This included studies to minimize surgical trauma and to ensure that the electrical stimulus parameters did not damage the spiral ganglion cells. The clinical trial of

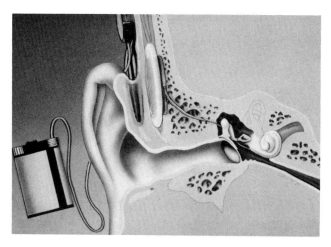

Fig. 1. Diagram of the Nucleus cochlear prosthesis.

the device was commenced in 1982 and premarket approval given by the FDA in 1985.

The Australian biomedical firm Nucleus Limited which developed the cochlear implant for clinical trial established subsidiaries Cochlear Pty Limited and Cochlear Corporation which are now marketing the device and undertaking further research and development.

A diagram showing the 22-electrode cochlear implant and speech processor can be seen in figure 1. The receiver-stimulator is implanted in the mastoid bone, and the 22 electrode array inserted around the scala tympani of the basal turn of the cochlea. The electrode is inserted through the round window or an opening drilled anterolateral to it, and access to the middle ear is obtained via a posterior tympanotomy. The patient wears a directional microphone above the ear and a transmitter coil placed on the skin directly over the underlying receiving coil. Coded speech information and power to operate the device are transmitted through the intact skin by radio waves. The speech sounds that are picked up by the directional microphone are processed by a speech processor which is a small unit that can be worn in a pocket or attached to a belt.

The clinical results obtained from the studies carried out at the centres using the feature-extraction multiple-electrode cochlear prosthesis have shown that it provides significant help to postlingually deaf patients in understanding running speech, especially when used in combination with lip-reading. About one third to one half of the patients can also understand some running speech without help from lip-reading.

Further research studies are continuing at the Department of Otolaryngology, University of Melbourne, to develop more advanced receiver-stimulators and speech processors with improved performances for both adults and children. Studies are also being undertaken to develop multiple-channel electrotactile and vibratory devices to help deaf patients communicate.

The Physiology and Psychophysics of Electrical Stimulation of the Auditory Nerve in the Experimental Animal

Before developing a multiple-electrode cochlear implant, physiological and behavioural studies were undertaken to determine how best to code frequency by electrical stimulation of the auditory nerve. These studies are listed in table 1. Adequate frequency coding with electrical stimulation is essential for a cochlear implant designed to restore speech understanding, as frequency carries the greatest amount of information.

There are two theories of frequency coding: the volley theory (time-period code), and the place theory. The volley theory, best referred to as a time-period code, postulates that pitch is perceived by decoding the time intervals between action potentials in the central auditory pathways. The second or place theory postulates that pitch is perceived by decoding the place or site of excitation in the brain. There have been many studies which have shown evidence in favour of both the timing and place theories. For example, the timing theory is supported by Wever and Bray [140] who showed that auditory responses followed the sound frequency and Rose et al. [117] who showed that the auditory nerve fibres of the squirrel monkey responded in phase to tones with frequencies up to 5.0 kHz, although each nerve fibre did not discharge every cycle. The place theory is supported by research by Galambos and Davis [81], Kiang [90] and Evans [73] who showed that auditory nerve fibres responded best to specific frequencies, and by Rose et al. [115, 116] and Tsuchitani and Boudreau [139] who found that the central auditory pathways were tonotopically organized so that a frequency scale was preserved.

Frequency Coding – Volley Theory or Time-Period Code

Physiology

In order to study how best to code frequency using a cochlear implant, an initial study was carried out by recording cell responses and field potentials from the superior olivary complex to electrical stimula-

Table 1. Studies on the physiology and psychophysics of electrical stimulation of the auditory nerve in the cat

---

1 Frequency coding – volley theory or time period code
  A Physiology
    Unit responses
    Field potentials
  B Psychophysics
    Rate discrimination – steady-state stimuli
    Rate discrimination – frequency-modulated stimuli
    Rate discrimination – signal detection theory
    Rate and intensity discrimination vs ganglion cell population
2 Frequency coding – place theory
  A Electrical resistance of cochlear structures
    Computer model
    Unit Responses
  B Current spread scala tympani – bipolar, common ground, monopolar stimulation
    Unit responses
    EABR masking study
    Tank studies
    Patient Study
3 Further frequency coding studies
  A Place coding
    Radial vs longitudinal stimulation
    EABR study with full- and half-bands
  B Time period code
    Unit responses

---

tion of the auditory nerve [26, 27]. It was considered appropriate to record from a level higher than the auditory nerve, for example the superior olive, as this would permit an evaluation of the effects of central processing mechanisms important in decoding frequency.

The results showed that responses could not be recorded at stimulus rates higher than 200–300 pulses/s. This was thought to be due to postsynaptic inhibition which could be as long as 20 ms in this nucleus [28]. In this case the inhibition would suppress the action potentials occurring close together in time at high stimulus rates. As only a small number of unit responses were sampled, field potentials from a population of cells were also recorded, and again the results showed that postsynaptic inhibitory mechanisms could interfere with the decoding of frequency on a timing basis.

Table 2. Detection of change in rate of electrical stimulation [29]

Rate pulses/s	100%		50%	
	apical	basal	apical	basal
100	*	*	n.s.	n.s.
200	*	*	*	n.s.
400	n.s.	n.s.	n.s.	n.s.

* Significant at 0.05 level; n.s. = not significant at 0.05 level.

As the data from the above studies had been recorded directly from auditory pathways in the anaesthetized cat, it could not be concluded that they were applicable to the decoding of frequency in the intact alert animal. Awake cats could perhaps learn to decode higher stimulus frequencies from information available in a single stimulus channel not detected in the experimental recording system used.

Psychophysics

To help resolve this matter a series of behavioral studies [29–31, 141] was undertaken on cats trained to respond to a change in rate of stimulation. Frequency difference limens for electrical stimulation were recorded and compared to those obtained with sound. It was considered that the just detectable differences in the perception of rate of electrical stimulation would indicate how well the animal would decode information in the speech frequency range, using a single-channel implant.

The results of the first study [29] showed difference limens for electrical stimulation that were only similar to those for acoustic stimulation up to a rate of 200 pulses/s (table 2). Table 2 shows that the cat was able to detect a change in stimulus rate of 100% for a rate of 100 pulses/s on both an apical and basal electrode. At 200 pulses/s the cat could detect a 100% rate change for both electrodes, and a 50% change for an apical electrode. At 400 pulses/s the cat could not detect a 100% change in rate. In this study the histological examination of the cochlea showed no hair cells present so the results were not due to electrophonic hearing. Electrophonic hearing occurs when an electrical pulse induces a mechanical vibration of the basilar membrane which in turn stimulates the hair cells. The results of this study showed the difference limens for electrical stimu-

Table 3. Rate (%) of electrical stimulation difference limens (frequency-modulated stimuli) [30]

Stimulus mode	Carrier frequency				
	200	400	800	1,600	3,200
Cat 1					
Acoustic	2.5	2.5	0.5	0.5	0.5
Electrophonic	0.5	0.5	0.5	25.0	25.0
Auditory nerve	25.0	25.0	25.0	>50.0	>50
Cat 2					
Acoustic	2.5	2.5	0.5	0.5	0.5
Electrophonic	5.0	5.0	25.0	25.0	50.0
Auditory nerve	5.0	25.0	25.0	50.0	>50.0

lation were not as good as those for acoustic stimulation. The frequency difference limens for acoustic stimulation in cats [80] and in humans [124] were 2–5% at 100 and 200 Hz.

The results from a second study [30, 31], where the difference limens were measured by presenting frequency modulated stimuli and where electrophonic hearing was controlled for by irradiating the cochlea, showed electrical difference limens similar to those for acoustic stimulation up to a rate of 200 pulses/s (table 3). From table 3 it can be seen that one cat had similar difference limens for sound and rate of stimulation of the auditory nerve at a carrier frequency of 200 pulses/s. Above this carrier frequency the difference limens were much poorer for electrical stimulation. The fact that some discrimination could be undertaken at rates of 400 pulses/s and above, was probably due to intensity differences produced when varying stimulus rates.

To further substantiate these animal behavioural results a third study was undertaken [141]. In this case signal detection theory was used to determine difference limens. Intensity differences with rate changes were controlled for, and hair cells were destroyed using neomycin. In this study an upper limit of 800 pulses/s was found for rate discrimination.

It is of interest that the results in animals for discrimination of rate of stimulation were similar to those obtained on cochlear implant patients [125, 130–132, 134, 137]. They also show that valuable information can be obtained from animal experimentation prior to cochlear implantation.

Not only did we examine the psychophysics of electrical stimulation on the cat with a normal auditory nerve population, but we considered it important to study how variations in the number of nerve fibres might affect the percepts. This would more closely model the situation in profoundly-totally deaf patients where diseases would have resulted in variable degrees of auditory nerve loss.

Experimentally deafened cats with differing populations of residual spiral ganglion cells were implanted and electrically stimulated. They were conditioned to respond to changes in electrical pulse rate and amplitude, and both electrical pulse rate and amplitude difference limens were determined. It was found that although there were some variations in difference limens between animals, the difference limens showed no correlation with residual spiral ganglion cell populations over the range 8–44% [11, 12].

This study showed, therefore, that a reduction in the population of ganglion cells and auditory nerve fibres that would occur in some profoundly-totally deaf patients should not affect their ability to discriminate amplitude or affect pitch coding on a timing or rate basis. Nevertheless, as electrical stimulation, regardless of the number of nerve fibres, has severe limitations in coding frequency it was important to study how to present pitch on a place basis.

Frequency Coding – Place Theory

Electrical Resistance of Cochlear Structures
To help provide pitch on a place basis, studies were undertaken to determine how best to localize the current to discrete groups of nerve fibres. Studies were undertaken to model the resistances of the cochlear structures, to determine the resistances to flow of current across the partitions of the cochlea, and also to determine the spread of current along the scala tympani of the basal turn. These investigations [7, 9, 10] were undertaken concurrently with those to determine the histopathological effects of electrode implantation at various anatomical sites [30–32, 34].

Our physiological studies [7] showed that if electrodes were placed so that stimulation occurred across the scalae, for example between the scala vestibuli and scala tympani, the high resistance of Reissner's membrane, in particular, would limit current flowing through the auditory nerve. As this method of stimulation would require electrodes to be placed directly into the scalae, and as our histopathological studies showed this could cause severe sensorineural damage, it was not pursued. The scala tympani insertion of electrodes, however, was shown to result in minimal damage to sensorineural elements, and for this reason studies were concentrated on how best to localize the current to discrete groups of nerve fibres using this surgical approach.

Current Spread in Scala tympani

The spread of current within the scala tympani of the cochlea was studied for stimulation using bipolar and monopolar electrodes as shown in figure 2. Using a technique developed by Merzenich [100] it was possible to determine the spread of current along the cochlea by recording the thresholds of cells in the inferior colliculus to the electrical stimulus. The inferior colliculus receives a binaural input so that the frequency of best response for an acoustic stimulus from the contralateral ear indicates the site of stimulation along the cochlea of the implanted ear. In this way the thresholds for electrical stimulation can be plotted at different points along the cochlea. The results in figure 3 show that for bipolar stimulation the voltage attenuation was 3–4 dB/mm, and for monopolar stimulation it was less than 1 dB/mm. These results were in agreement with those recorded by Merzenich [100] who showed attenuation which ranged between 3.3 and 8 dB/mm. In this study by Merzenich high attenuation only occurred close to the stimulating electrode, and was partly due to the fact that a moulded array was used which filled a large proportion of the scala tympani and therefore restricted the flow of current along it.

The results from our study described above were obtained with a pair of bared wire electrodes lying radially across the scala tympani. We have also carried out a study in cats with a banded free fitting array similar to that used in humans. In this study [108] the spread of current was determined by recording the EABR for a reference stimulus at one location in the cochlea, and measuring the extent to which a test stimulus at another location resulted in the suppression of the EABR from the reference stimulus, which was presented within the refractory period of the test stimulus. In this case an attenuation of 2–3 dB/mm was obtained, and this was similar to that in our previous study reported above.

As the dynamic range of an electrical stimulus from threshold to maximum discomfort level varies from 6 to 12 dB [134] the spread of current along the cochlea at a maximum stimulus level will be about 2–4 mm on either side of the stimulating electrode. As this current spread occurs at

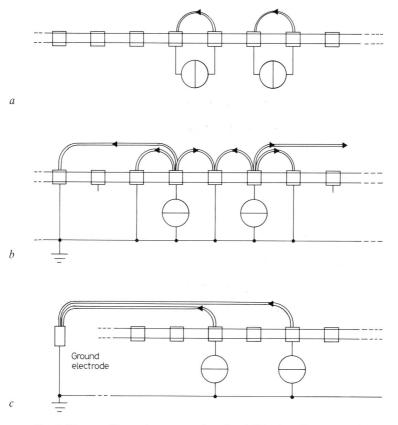

Fig. 2. Diagram illustrating current flow for *(a)* bipolar, *(b)* common intracochlear ground and *(c)* monopolar stimulation.

a maximum level it will be less at a comfortable listening level, and will be sufficiently localized to allow pitch to be conveyed on a place basis. The Nucleus electrode array has a spatial separation of 0.75 mm between each electrode, and as will be discussed below, the patients were able to scale pitch on a place basis on all electrodes when a stimulus was presented at a comfortable listening level.

Although bipolar stimulation is standard for the Nucleus multiple-electrode receiver-stimulator this could not be carried out in the proto-

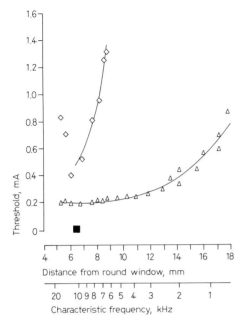

Fig. 3. The threshold of inferior colliculus units to electrical stimulation from an electrode situated 6–7 mm from the round window, plotted as a function of the characteristic frequency of the units and their computed distances from the round window. Results for both monopolar △ and bipolar ◇ stimulation of the scala tympani are shown. The approximate position of the stimulating electrode is shown by the position of the black square [50].

type receiver-stimulator first implanted by the University of Melbourne in 1978 for electronic reasons. To isolate the current to discrete groups of fibres, pseudobipolar or common ground stimulation was employed as shown in figure 2. With common ground stimulation the current from an active electrode passes to a ground in the cochlea, and is restricted to groups of nerve fibres. As shown in tank studies and from results on patients, current was adequately localized by common ground stimulation [10, 134].

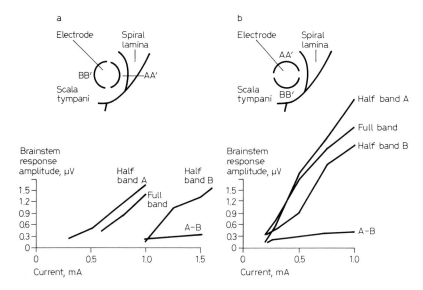

Fig. 4. Brainstem response amplitude growth curves for full- (A and B) and half-band (A or B) electrodes in two positions (a and b) varying by 90 degrees of rotation [44].

Further Frequency Coding Studies

Place Coding

Although our research has shown that current localization with the free-fitting banded electrode is satisfactory for 22 channels of stimulation, we have carried out an investigation to see if current localization could be improved with radial rather than longitudinal stimulation using a half-band instead of a full-band electrode. If current localization could be improved it would allow an increase in the number of stimulus channels that could be used to convey place pitch.

The stimulus configurations for half-band and full-band electrodes using radial and longitudinal stimulation are shown in figure 4. The stimulus characteristics of the electrode configurations were assessed by recording the amplitude of the brainstem responses for different current intensities. An example of the results obtained can be seen in figure 4 [42–44]. Figure 4 shows the amplitude growth curves for half-band electrodes placed as shown in the inset. During longitudinal stimulation between the two half-band electrodes, A–A¹, close to the spiral lamina, the threshold was lower than for the two half-band electrodes, B–B¹, further from the spiral lamina. Furthermore, the results obtained during longitudinal stimulation between the two full-band electrodes (A and B, and A¹ and B¹ joined together) lay between those for the two half-band electrodes. During radial stimulation (that is between A and B), the threshold was high and the growth of response with current was small. These latter findings imply that radial stimulation between half-band electrodes would provide improved current localization.

After making the recordings in the positions shown in figure 4 the electrode array was then rotated through 90 degrees as shown in the inset in figure 4, and the thresholds and amplitude growth responses measured again. This resulted in a change in the thresholds and slope of the growth curves. The full-band growth curve had changed significantly and had a lower threshold indicating the electrode had moved closer to the nerve fibres. The results showed less difference between longitudinal stimulation between the two pairs of half-band electrodes A and B. The threshold for radial stimulation dropped significantly as would be expected if the electrode had moved closer to the nerve fibres, and its growth response function was still very small.

The results in the study confirm the findings of Merzenich and White [101] that current localization is better with radial rather than longitudinal stimulation between electrode pairs. The stimulus thresholds, however, varied depending on whether the electrode lay close to or at a distance from the nerve fibres. This variation in threshold makes radial stimulation more difficult with a free-fitting electrode. Furthermore, longitudinal stimulation between pairs of half-band electrodes did not provide any real advantage in current localization compared to full-band stimulation. The slight advantage seen is more than offset by the fact that during insertion into the human cochlea there is rotation of the electrode as it passes around the basal turn. Furthermore, rotation may also be carried out by the surgeon to facilitate its insertion, and disease and anatomical variations will prevent accurately locating the electrode array to take advantage of either half-band longitudinal or half-band radial stimulation. Consequently, these electrophysiological findings together with the biological safety advantages of a free-fitting electrode [128] make the Nucleus banded electrode array an optimal solution for multiple-electrode intracochlear stimulation.

Not only was it clear from the studies outlined above that bipolar or

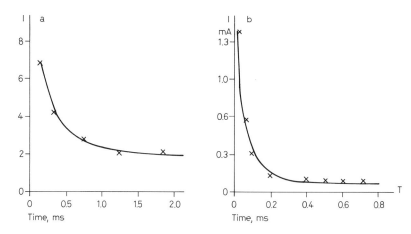

Fig. 5. a Strength-duration curve from a frog sciatic nerve. *b* Strength-duration curve from a cat auditory nerve [33, 35].

common ground stimulation should be used in the design of a receiver-stimulator, but other stimulus parameters needed to be defined. As mentioned under biological safety, pulses need to be biphasic so that charge can be balanced between phases to minimize corrosion and the release of toxic products. Furthermore, the width of each pulse needed to be defined so that an adequate charge could be delivered to excite the nerve fibres. This could be determined from a strength-duration curve, in which the current is plotted against pulse duration for a threshold response. A strength-duration curve for the auditory nerve is shown in figure 5 [33, 35]. From this it can be see that the pulse width should not be shorter than 0.1 ms otherwise the current required to reach threshold rises steeply, and would be impracticable with present technology. Finally, constant current stimulation rather than constant voltage stimulation is required as current rather than voltage is responsible for neural excitation. The stimulus current may vary unpredictably with variations in impedance unless constant current stimulation is used.

The electrophysiological and psychophysical studies discussed above were important in the design of the prototype receiver-stimulator and ultimately the clinical trial version manufactured by Nucleus Limited. Recently we have undertaken studies to examine in more detail how the auditory system responds to electrical stimulation so that the design of more advanced third generation devices may be possible [88].

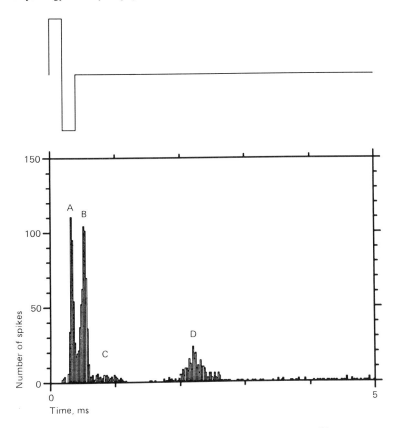

Fig. 6. Composite period histogram showing auditory nerve fibre responses to biphasic electrical current pulses. Responses were categorized into four types (A–D) on the basis of latency [88].

Time-Period Code

The first study has involved recording auditory nerve responses for a number of different intensities. A composite period histogram for auditory nerve fibre responses to a biphasic current pulse in normal hearing cats is shown in figure 6. This composite histogram illustrates the characteristics of the animal model. At the top you can see one period of the electrical stimulus, and at the bottom the time structure of the response. The response has four distinct components. We have labelled these A, B, C and D based on their latencies. These four components do not all occur simultaneously but this depends on the intensity of the stimulation.

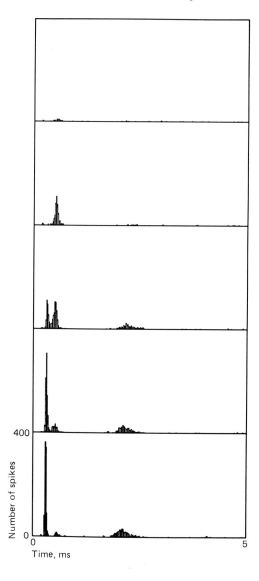

Fig. 7. Period histograms in response to 200 μs/phase pulses presented at five different intensities increasing from the top to the bottom. Intensities are in dB re 1 μA [88].

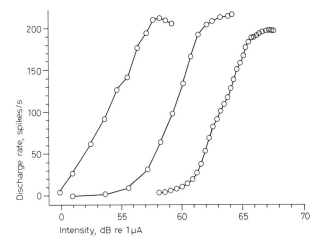

Fig. 8. Rate-intensity functions for three fibres, in response to 200 µs/phase pulses presented at 200 pulses/s in 100 ms bursts [88].

The A and B responses were highly synchronized to the current pulse, and they occurred at latencies of about 0.3 and 0.6 ms after the pulse onset. This suggests that the 0.3-ms responses were generated at the axon and the 0.6-mm responses at the dendrite. The C response is probably a poorly synchronized version of the B response. The D response occurred after at least 1.5 ms. This long latency implies that the D response was not induced by direct electrical stimulation, but almost certainly arose electrophonically.

The patterns of auditory nerve responses to increases in intensity are shown in figure 7. This figure shows period histograms for different intensities. At threshold the stimulus elicited only the B or dendritic response. The B response increased in size with intensity as the fibres fired more repeatedly with each stimulus. At supra-threshold levels A and D responses arose. At higher intensities the A response grew rapidly at the expense of the B response, and was very highly synchronized. At the highest intensities the B response virtually disappeared, leaving only the A response and the D response. These findings show that our auditory nerve data are dominated by short-latency responses with little acoustic contamination.

Input-output functions for intensity were also recorded to see whether there was a correlation with psychophysical loudness growth functions in implanted patients. These are shown in figure 8 for a stimulus

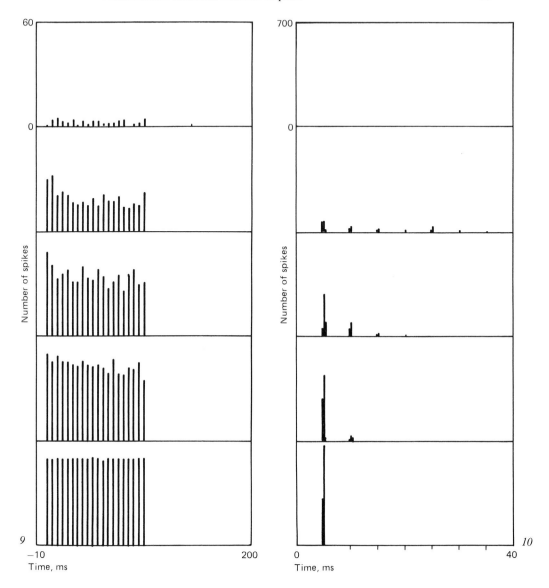

Fig. 9. Poststimulus time histograms obtained at various intensities of electrical stimulation increasing from the top to the bottom in response to 200 μs/phase pulse trains presented at 200 pulses/s for 100 ms. Intensities are in dB 1 μA [88].

Fig. 10. Interspike interval histograms obtained at various intensities of electrical stimulation increasing from the top to the bottom in response to 200 μs/phase pulse trains presented at 200 pulses/s for 100 ms. Intensities are in dB re 1 μA [88].

of 200 pulses/s. The input-output functions for three fibres with low, medium and high thresholds are shown. The firing rate increased rapidly until it reached a plateau at 200 spikes/s. The range of intensities over which the response grew from threshold to a maximum, or its dynamic range, was found to be narrow and in this case 5 dB. This narrow dynamic range for intensity in electrical stimulation is also seen in the psychophysical results on implant patients.

For the coding of speech it is not only important to study the response of fibres to the intensity of an electrical pulse, but also the fibre responses to the frequency and timing of the pulses. This was undertaken by recording poststimulus-time and inter-spike-interval histograms.

The poststimulus-time histograms for electrical pulse trains are shown in figure 9. The poststimulus-time histograms in response to electrical stimuli are not very different from those obtained in response to acoustic tones. The initial peak decays to a steady-state value and the steady-state response is maintained as long as the stimulus is present. The main difference is that at high current levels every pulse becomes capable of eliciting a spike which produces a rectangular histogram.

The inter-spike-interval histograms for electrical stimulation are shown in figure 10. Here we see that at lower intensities there is some similarity between acoustic and electrical stimulation as the pulses occur at regular multiples of the stimulus period. However, at high intensities all of the inter-spike intervals are at the electrical pulse rate, even up to 600–800 pulses per second. The highly synchronized response to electrical stimulation and the lack of spikes distributed over a range of intervals may be the reason frequency discrimination is much better for acoustic stimulation than it is for electrical stimulation, seen with the psychophysical results in animals and the psychophysical results in implant patients presented below.

Although there is a correlation between the physiological results in animals and the psychophysical results in our implant patients, considerable more research will be required to more fully understand the underlying coding mechanisms involved with electrical stimulation so that improved speech processors can be developed.

Biological Safety

Biological safety is an important issue for cochlear implantation as we do not want to damage the auditory nerve fibres it is hoped to stimulate electrically. The investigations carried out are listed in table 4 and were in the following areas: the biocompatibility of materials; the development of an atraumatic surgical approach; the prevention of an infective labyrinthitis postimplantation, and safe electrical stimulus parameters.

The Biocompatibility of Materials

Initial Studies

It was considered important to assess the biocompatibility of possible cochlear implant electrode materials as they can vary in their tissue toxicity [22]. This information was also important in interpreting the histopathological effects of intracochlear electrode implantations and electrical stimulation. As the medical grade silicone elastomers have been implanted for many years with minimal inflammatory reactions [22, 107], these compounds were the ones most intensively examined in the present study. Other candidate materials for the electrode carrier were assessed. These were polytetrafluorethylene (PTFE), PTFE sputtered onto medical grade Silastic® tubing, hexafluoropropylene (FEP), and polyurethane. Platinum 99.9% electrodes are satisfactory from an electrochemical point of view [24], and are used extensively in neural prostheses. Prior to its final selection the biocompatibility of platinum was also evaluated. The initial studies on the biocompatibility of materials are listed in table 5.

The toxicity of the materials was first determined by implanting them subcutaneously and intramuscularly in both rats and cats. This allowed their inflammatory responses to be compared with known reactive and non-reactive controls. The experiments also provided baseline tissue

Table 4. Studies on biological safety

1 Biocompatibility of materials
2 Development of an atraumatic surgical approach
3 Prevention of infective labyrinthitis postimplantation
4 Safe electrical stimulus parameters

Table 5. Studies on the biocompatibility of materials

1 Initial experimental studies – candidate materials
   A Normative data – rats
      Reactive and non-reactive controls
      Location and duration of implantation
   B Candidate materials – rats
      Subcutaneous
      Intramuscular
   C Candidate materials – cats
      Intramuscular
      Intracochlear
   D Controls for biological variables – cats
      Intracochlear
2 Further Studies for FDA – materials in Cochlear Pty Limited receiver-stimulator and electrode array
   A Cytopathic effect – embryonic cells
   B Systemic toxicity – mice
      Intravenous
      Intraperitoneal
   C Intracutaneous irritation – rabbits
   D Ninety-day implant tests – rabbits
      Subcutaneous
      Intramuscular
   E Test of assembled units – cats
      Intramuscular

responses that could be compared with those obtained from the implantation of the materials in the cochleas of cats.

General Methods

The methods used in general for assessing the toxicity of the biomaterials was a slightly modified version of the recommendation of the US Pharmacopeia [148, 149]. Albino rats were used instead of albino New Zealand rabbits as they are more robust and readily available. A series of implants were also carried out on cats to make sure there were no species differences that would affect the interpretation of the results on cat cochleas. Small incisions were made through the skin, and the implants inserted into the subcutaneous tissues and paravertebral muscles under vision, rather than introducing them blind through the skin. It was considered important to only report on the tissue reactions close to the material. The implants were left in situ for 14 rather than 7 days so the healing processes would be well established, and any acute reaction would therefore be a response to the implanted material rather than the surgery. A reactive, as well as a non-reactive control was also used. Medical grade Silastic and PTFE were the non-reactive controls, and as pilot studies showed that the Silastic 382 catalyst produced a marked inflammatory response at 14 days, this was injected as the reactive control. Four to six animals were used for each test material rather than the recommended two, in order to better assess variations in responses between animals. When documenting the response a macroscopic classification was not used as it is not a sensitive method. A microscopic grading was carried out in all cases. Furthermore, reports were only made when an implant pocket could be demonstrated. The response around the implanted site was classified into six grades of severity for polymorphonuclear leucocyte, mononuclear leucocyte, and fibrous tissue reactions. If there were any giant cells, foreign body debris, fatty infiltration or necrosis they were noted. The classification was carried out by a histopathologist with experience in inflammatory research. Furthermore, in reporting on the tissue reactions he was given no information on the sections and the materials used [G.B. Ryan].

Study 1

Methods. The first study was a preliminary investigation on four rats to evaluate the variation in responses to a reference non-reactive material (medical grade Silastic) at different locations in the one animal, and to compare the tissue responses for implant durations of 14 and 56 days. It also served to measure test-retest reliability and to establish reference data for the non-reactive material, medical grade Silastic. The responses to the Silastic 382 catalyst were recorded as baseline data for the reactive control.

The materials were washed in soapy water followed by 10 rinses with distilled water before autoclaving. The Silastic 382 catalyst was also autoclaved. The animals were anaesthetized with diethyl ether, and the implants made using a clean technique. After small incisions in the skin, the materials were placed in subcutaneous pockets and the wounds closed. Medical grade Silastic tubing was placed at four sites in four rats, and the Silastic 382 catalyst injected at one site in each animal. Two rats were sacrificed in 14 days and the other two in 56 days. The tissue was cut to find the implant material and tissue pocket, and then embedded in paraffin wax, sectioned at a thickness of 8 μm, and every tenth one stained with haematoxylin and eosin.

Results. The tissue inflammatory responses for the non-reactive control (medical grade Silastic), and the reactive control (Silastic 382 catalyst) in study 1 were classified at high magnifications to be sure of the cell types. For polymorphonuclear and mononuclear leucocytes the grading was: grade 1 (very slight) – occasional cells in the high power fields; grade 2 (mild) – a few cells in most fields; grade 3 (moderate) – a few clusters of cells in some fields; grade 4 (moderate-severe) – frequent clusters of cells in many fields; grade 5 (severe) – large numbers of cells in some fields; grade 6 (very severe) – large numbers of cells in most fields. The fibrous tissue was graded from 1 to 6 on the thickness of the tissue. The maturity of the fibrous tissue was also noted.

The results showed that for medical grade Silastic the implant site and animal did not contribute significantly to the classification of each type of tissue response. This applied for implantations of 14 and 56 days. It is also worth noting that there was little difference in the polymorphonuclear leucocyte and fibrous tissue responses for the two implant durations, but the mononuclear leucocytes were less prevalent with the 56-day implants. The Silastic 382 catalyst resulted in a vigorous 5–6 level response for mononuclear leucocytes and fibrous tissue, and a 2-level one for polymorphonuclear leucocytes. These results confirm its suitability as a reactive control.

Study 2

Methods. The second study was undertaken to examine the tissue reactions obtained with a number of materials in two sets. In one set there were: Silastic MDX-4-4210, Silastic medical adhesive type A, and platinum 99.9% sheet with a polished surface. In the other set there were Silastic 382, FEP, PTFE sputtered onto medical grade Silastic, and polyurethane. Medical grade Silastic was used as the non-reactive control, and the Silastic 382 catalyst as the reactive control.

The materials were all cleaned with ultrasonic agitation in Freon for 10 min, in 100% propane-2-ol for 10 min, and in two changes of distilled water for 10 min each. They were then stored in distilled water until ready for autoclaving. The Silastic 382 catalyst was drawn into a sterile syringe prior to injection.

The materials in each set were implanted aseptically in five rats into subcutaneous or, where possible, intramuscular pockets through small incisions in the skin on the back. The animals were anaesthetized with Ketamine (18 mg i.m.i.) and Xylazine® (3.8 mg i.m.i.). After 14 days they were sacrificed and transcordially perfused with an isotonic phosphate buffer containing heparin, followed by buffered Karnovsky's fixative. A large area of skin, subcutaneous tissue and muscle was removed from the back, submerged in fixative, and each implanted material dissected out with 5 mm of tissue surrounding it. The implanted tissue was dehydrated and embedded in Spurr's low-viscosity resin, and serially sectioned at a thickness of 2 µm. Sections were stained with haematoxylin and eosin, and the inflammatory responses graded.

Results. The results of the second study on the biocompatibility of various candidate materials (Silastic MDX-4-4210; Silastic medical adhesive type A; Silastic 382; FEP; PTFE sputtered onto medical grade Silastic; polyurethane, and platinum 99.9% sheet) implanted subcutaneously and intramuscularly showed there were variations in the polymorphonuclear responses to the candidate materials at 14 days. The most consistently low polymorphonuclear responses were for PTFE and polyurethane. The responses to Silastic MDX-4-4210 and Silastic medical adhesive type A were minimal. The mononuclear leucocytes and fibrous tissue responses were greatest for Silastic 382 and PTFE sputtered onto medical grade Silastic suggesting that these were unsuitable as electrode carrier materials.

Study 3
Methods. In the third study the following materials were all implanted into the paravertebral muscles of six cats: Silastic MDX-4-4210 clean grade elastomer, medical grade Silastic, PTFE sputtered onto medical grade Silastic, FEP, polyethylene and polyurethane. Silastic medical grade tube and PTFE were implanted as non-reactive controls. In the same animals the following candidate electrode carrier materials were also implanted in the cochlea: Silastic MDX-4-4210 clean grade elastomer

(2 cochleas), FEP (3), polyethylene (2), polyurethane (3) and PTFE sputtered onto Silastic medical grade tubing (2).

The cats were anaesthetized with Ketamine (18 mg/kg i.m.i.) and Xylazine (3.8 mg/kg i.m.i.). In each cochlea the round window membrane was incised and a 3–4 mm length of material placed in the scala tympani so that the terminal end was flush with or under the membrane. A 0.5-ml solution of ampicillin 0.1% and cloxacillin 0.05% was instilled into the middle ear. Ampicillin (200 mg, i.m.i.) was administered during the surgery and continued orally for 7 days. The animals were sacrificed after 119–129 days. They were deeply anaesthetized, the bullae were opened and flooded with buffered Karnovsky's fixative, and the animals perfused arterially with heparinized 0.1 M cacodylate buffer in normal saline, followed by 0.1 M cacodylate buffered Karnovsky's fixative. The otic capsules were thinned and the cochleas decalcified in ethylenediaminetetraacetic acid (EDTA) before finally embedding in Spurr's low viscosity resin. The cochleas were serially sectioned at a thickness of 3 μm, and sections every 100–130 μm stained with haematoxylin and eosin.

Results. In the third study where some of the candidate electrode materials (Silastic MDX-4-4210; medical grade Silastic; PTFE sputtered onto medical grade Silastic; FEP; polyethylene, and polyurethane) were implanted intramuscularly in cats, the results showed that after 119–129 days there was little polymorphonuclear leucocyte response to all the materials, and that the fibrous tissue was usually classified at level 2. On the other hand, there was a greater variation in the mononuclear leucocyte responses. These were greater for PTFE sputtered onto medical grade Silastic, polyurethane and FEP. The responses were least for medical grade Silastic, Silastic MDX-4-4210, and PTFE. Although the numbers are small the results suggest that medical and clean grade silicone rubbers as well as PTFE are the least reactive.

In the third study, materials were also placed inside the scala tympani of the basal turn of the cat cochleas. Cochleas with implanted Silastic MDX-4-4210, PTFE sputtered onto medical grade Silastic, FEP, polyethylene and polyurethane were available for histological investigation. The tissue responses were graded in a similar way to that used for intramuscular implants. The results showed that there were minimal responses to all materials except when trauma and infection were present.

In two cochleas there were reduced spiral ganglion cell numbers in the basal turn. These cochleas had implants of PTFE sputtered onto medical grade Silastic and FEP. Calcification occurred in only one of the cochleas and was not due to infection. In this case polyurethane was the implanted material. The polyurethane tube had a larger diameter than the other implants, and this could have been an important factor in causing injury to the endosteal lining.

Study 4

Methods. The fourth study was undertaken on the effect of blood within the scala tympani as a biological variable in the previous study on cat cochleas. In two cats 0.02 ml of sterile normal saline were injected into the cochlea through the round window membrane on one side, and 0.02 ml of the animal's blood on the other side. 0.5 ml of a solution of ampicillin 0.1% and cloxacillin 0.05% were then injected into the middle ear remote from the round window. Finally, in one animal the implantation of medical grade Silastic and the injection of normal saline was compared with the implantation of medical grade Silastic and the injection of 0.02 ml of the animal's blood. Ampicillin (200 mg, s.c.) and cloxacillin (200 mg, s.c.) were administered during the surgery, and ampicillin continued orally for 5 days. The cats were sacrificed, and their cochleas embedded, sectioned and stained as described above.

The animals in all our studies were housed and managed according to the National Health and Medical Research Council of Australia's 'Code of Practice for Control of Experiments in Animals', and the US Food and Drug Administration's 'Good Laboratory Practice' guidelines.

Results. The fourth or control study showed no cochlear tissue responses or loss of spiral ganglion cells and dendrites, following the injection of saline and blood into the cochlea. This means that blood within the scala tympani was not a factor leading to histopathological changes such as calcification.

Discussion

Considerable work has been undertaken on the biocompatibility of materials with special reference to bone, joint, and heart valve replacements [142]. Their biocompatibility, however, varies with the implant site and surrounding tissues [22]. Consequently, materials used for cochlear implant electrodes should be evaluated in vivo in animal cochleas. Furthermore, as the reports in the literature on the tissue responses for individual materials vary, and as the composition of the materials themselves can differ, it was considered essential to also assess the biocompatibility of the actual materials likely to be used by subcutaneous and intramuscular implantations.

The materials used should also have the appropriate mechanical properties which include a low coefficient of friction and a stiffness which allows the electrode tip to be advanced during the insertion of the electrode without causing trauma. The carrier should also be easy to mould and sterilize, and not be permeable to body fluids. No materials have all these attributes, but Silastic MDX-4-4210, Silastic medical adhesive type A, and medical grade Silastic were considered to have the most potential.

The results of the subcutaneous and intramuscular implants on rats and cats showed firstly that Silastic MDX-4-4210 is biocompatible and induces little inflammation. The platinum electrode also induces little tissue response. Other materials that could be suitable for implantation are medical grade Silastic, PTFE, FEP, and polyethylene. The results indicated some doubt, however, with Silastic 382, and PTFE sputtered onto Silastic. The catalyst for Silastic 382 produces a strong tissue reaction and could diffuse out of the material over time leading to continuing tissue toxicity. The sputtering process will break the molecular chains in PTFE and lead to the production of toxic products.

The results for the subcutaneous and intramuscular implants on rats and cats showed good correlation across the two species, and in general similar tissue responses were obtained for subcutaneous and intramuscular implants. Intramuscular implants can, however, have variations in the inflammatory response around the implant due to the direct effect of muscle pull, and this must always be taken into consideration when interpreting results.

Although the number of intracochlear implants was small, it was found that providing infection or trauma did not arise there was good correspondence with the subcutaneous and intramuscular findings. Again, Silastic MDX-4-4210 and FEP were not strongly reactive. It was interesting that there was little inflammatory response to the PTFE sputtered onto Silastic but a localized loss of spiral ganglion cells which could have been due to the release of toxic products produced as a result of the sputtering process. On the other hand there was quite a marked chronic inflammation with bone formation around the polyurethane electrode carrier, but no loss of spiral ganglion cells. As the polyurethane lay against the spiral lamina and basilar membrane, the inflammatory response was probably due to trauma and microfractures of the spiral lamina.

Conclusion

The study showed that Silastic MDX-4-4210, Silastic medical adhesive type A, and platinum 99.9% sheet are biocompatible and produce a minimal tissue response. PTFE, FEP and polyethylene were also shown to be biocompatible, but not PTFE sputtered onto Silastic or Silastic 382. There was good correspondence between the tissue reaction to subcutaneous and intramuscular implantation and between rats and cats. The results with intracochlear implantations were similar to those obtained in subcutaneous tissue and muscle.

These studies supported the choice of Silastic MDX-4-4210, Silastic medical adhesive type A and platinum 99.9% for the intracochlear electrode array.

Further Studies

Further biocompatibility studies were undertaken for the Premarket Approval Application to the FDA on the materials used to construct the Nucleus electrode array, lead wire assembly and receiver-stimulator, that would be in contact with the body tissues. These materials are: Silastic medical adhesive type A; Silastic tubing 602; Silastic MDX-4-4515; Silastic MDX-4-4210; platinum band 99.9% pure. The biocompatibility studies are listed in table 5.

Cytopathic Effects

Silastic medical adhesive type A, Silastic tubing 602, Silastic MDX-4-4515 and Silastic MDX-4-4210 were evaluated for a cytopathic effect by placing the material in direct contact with a confluent monolayer of human embryonic cells. After an incubation period of 24 h, the cytopathic effect was evaluated microscopically and compared with a positive and negative control. No cytopathic effect was produced by the test materials.

Systemic Toxicity

Systemic toxicity was evaluated in mice by a method described in the US Pharmacopeia [150] on test materials. Silastic medical adhesive type A, Silastic tubing 602, Silastic MDX-4-4515 and Silastic MDX-4-4210 were extracted with sterile 0.9% saline solution and cotton seed oil. Saline extracts were injected intravenously into a group of five mice at a dosage of 50 ml/kg, while a second group of five mice was injected intraperitoneally with cotton seed extract at a dosage of 50 ml/kg. The third and fourth groups of mice were each injected with the extracting medium (control) in the same manner. The animals were observed immediately after injection, 1 and 4 h later, and after 1, 2 and 3 days. No mortalities, untoward behavioural reactions, or body weight losses were observed in any of the animals during the 3-day observation period for the saline, cotton seed oil and control extracts. Results indicate that the test materials did not contain any leachable components, in saline or cotton seed oil which were toxic when injected systemically.

Intracutaneous Irritation Tests

Intracutaneous tests were conducted on rabbits to evaluate intracutaneous irritation properties of saline and cotton seed extracts of Silastic medical grade adhesive type A, Silastic tubing 602, Silastic MDX-4-4515 and Silastic MDX-4-4210. The test materials were extracted with 0.9% saline solution and cotton seed oil. Each extract of the material was injected (0.2 ml) intracutaneously into 10 sites on the left side of the back of each of two animals, while an equal volume of the extracting medium (control) was injected into 10 sites on the right side. After injection the sites were examined for erythema and oedema at 1, 2 and 3 days. Neither erythema nor oedema were noted at any of the test sites at any of the three evaluation periods. Results indicate that the materials tested do not contain, when extracted with saline or cotton seed oil, any leachable constituents which are irritating when injected intracutaneously.

Ninety-Day Implant Tests

Ninety-day implant tests were conducted to evaluate the in vivo tissue response over time to materials implanted into the paravertebral muscles and ventral subdermal areas of male rabbits. Three groups of two animals each were implanted with three samples 15 mm apart from each other into the paravertebral muscles of each rabbit. In a similar fashion two strips of control polyethylene were implanted in opposite sides of each animal. After an observation period of 10, 30 and 90 days the animals were sacrificed and sites were examined grossly and microscopically. No differences were observed grossly or microscopically at necropsy between the control material and Silastic tubing 602, Silastic MDX-4-4515 and Silastic 4-4210. Microscopically the Silastic medical grade adhesive type A showed changes indicating that it is a slight irritant. The results from the above in vitro and in vivo studies demonstrated that Silastic tubing 602, Silastic MDX-4-4515 and MDX 4-4210 are not toxic or irritating. Silastic medical adhesive type A was found to be non-toxic and a slight irritant but not to an extent that it would prevent use of this material in the device.

Test of Assembled Units

In this study five Nucleus-assembled receiver-stimulator units were implanted intramuscularly in cats for a period of approximately 4 weeks. On removal of the units biopsies were taken from tissue adjacent to the units. Each biopsy was examined and photographed under the microscope. The degree of tissue reaction was recorded on the basis of the relative presence of polymorphonuclear leucocytes, mononuclear leucocytes and the amount of fibrous tissue. The total implantation time was 39 days in two cats, 34 days in one cat and 26 days in two cats. The appearance of the fibrous tissue capsule was moderate to fine. Mononuclear leucocytes were present in almost all cases, but polymorphonuclear leucocytes were absent. Foreign body giant cells were absent. The results obtained in this study suggest that the device will not elicit an adverse tissue response.

99.9% Pure Platinum Band

Finally, 99.9% pure platinum is well established as biocompatible. Extensive documentation demonstrates the acceptability of platinum as an electrode material for neural stimulation and this is summarized by Hambrecht [85] and Brummer and Turner [25]. Platinum has been shown to be biocompatible with neural tissue by numerous investigators and has been the electrode material of choice in a variety of numerous prosthetic applications.

An Atraumatic Surgical Approach

Having established that the materials we were using were biocompatible we also carried out a number of studies to develop an atraumatic surgical approach for multiple-electrode cochlear implantation. This has involved experimental studies on animals and human temporal bones and measuring the mechanical properties of electrodes. The studies are listed in table 6.

Table 6. An atraumatic surgical approach

1 Extracochlear implantation
   A Current spread
   B Histopathology
2 Intracochlear implantation
   A Histopathology
   B Electrode insertion in human temporal bones
      Histology
      Surface preparation
   C Electrode mechanical properties
   D Anatomical studies
   E Explantation and reinsertion

Extracochlear Implantation

We have carried out a preliminary study to see if multiple-electrode stimulation could be effectively carried out by placing electrodes in or on the bone overlying the cochlea, rather than use an intracochlear approach. If multiple-electrode extracochlear stimulation could be carried out without entering the cochlear it should be less traumatic. On the other hand, bone has a high electrical resistance [95, 113], and for this reason extracochlear stimulation should require high current levels for adequately stimulating discrete groups of auditory nerves.

To examine this matter, we carried out a preliminary study and drilled holes at different depths through the bone overlying the cat cochlea. Using extracochlear bipolar stimulation between electrodes placed in these holes we recorded the electrically evoked auditory brainstem response (EABR) thresholds and growth functions. It was considered that firstly the thresholds would depend on the resistance of the bone to the stimulating current, and secondly the slope of the growth functions on the spread of current to the auditory nerve fibres. A greater spread of current is likely to result in a more rapid recruitment of excited fibres with increases in intensity, and consequently a steeper slope in the input-output function. This difference is seen for radial versus logitudinal stimulation of the cochlea where the current spread is less for radial compared to longitudinal stimulation [44].

The results are shown in figure 11. The holes marked 1–4 were drilled at different depths in sequence along the basal turn of the cochlea. Hole

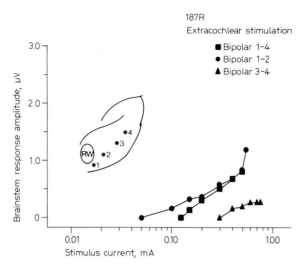

Fig. 11. Diagram of the electrically evoked auditory brainstem responses for bipolar stimulation between electrodes placed in holes 1–4 drilled at different depths through the bone overlying the cochlea. 1 = in perilymph of scala tympani; 2 = 0.2 mm superficial to scala tympani; 3 = 1.0 mm superficial to scala tympani; 4 = 2.0 mm superficial to scala tympani.

No. 1 was actually in the perilymph. When stimulating between sites 3 and 4, an EABR due to direct stimulation of auditory nerve fibres could not be produced at the maximum output of the stimulator, but an electrophonic response could be recorded. This finding supports the view that the impedance of the bone is too high to permit localized stimulation of groups of auditory nerve fibres at low and moderate stimulus currents. The other input-output functions were between sites 1 and 2, and 1 and 4. In these cases brainstem responses were recorded at lower thresholds, as the electrode at site 1 was actually in the perilymph and therefore had a lower impedance. Direct electrical stimulation of the auditory nerve fibres was possible at the upper limit of the stimulator output. It also appeared that the growth function of the EABR would be steep indicating a rapid recruitment of fibres due to poor current localization, but there were too few data points to be sure due to the limit on the stimulator output.

We also recorded the EABR input-output function for monopolar stimulation. This was carried out between two sites overlying the cochlea

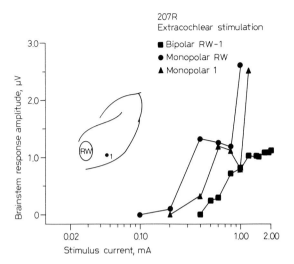

Fig. 12. Diagram of the electrically evoked auditory brainstem responses for monopolar stimulation between the sites No. 1 over the basal turn of the cochlea and the round window, and a common ground. The input-output function for bipolar stimulation between site No. 1 and the round window is also shown. Site No. 1 – drilled to endosteum.

and a remote ground. These results were compared with those for bipolar stimulation between an electrode site overlying the cochlea and the round window membrane. The results are shown in figure 12. From this it can be seen that the two monopolar stimuli gave similar growth functions and they were steeper than for the bipolar stimulus. These results suggest that monopolar stimulation between active sites over the cochlea and a common ground recruit fibres too rapidly for it to be effective in localizing current to discrete groups of fibres, and therefore would not be suitable for multiple-electrode stimulation.

In summary, these experimental studies showed that multiple-electrode stimulation is not likely to be effective with extracochlear electrodes. Furthermore, histological examination of the cochleas of cats up to 22 days after the bone had been drilled down to the endosteal lining showed that tissue damage in the cochlea could occur (fig. 13). Fibrous tissue and bone formation could also occur beneath the electrode, which would then raise the threshold to electrical stimulation over a period of time.

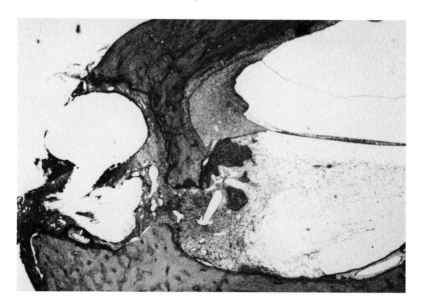

Fig. 13. Photomicrograph of the basal turn of the cat cochlea 22 days following implantation with an electrode placed in a hole drilled down to the endosteal lining of the scala tympani. ×450.

Intracochlear Implantation

As effective multiple-electrode stimulation would be unlikely with extracochlear electrodes an important question was: Could electrodes be placed intracochlearly without damaging the auditory nerves? Furthermore, should we place the electrodes directly into the cochlea by drilling through the bone at a number of locations, or should we insert a bundle of electrodes around the scala tympani through a single opening in the round window or the bone nearby?

Histopathology – Cats

To help answer these questions we implanted electrodes in the cat by either drilling directly into the cochlea or passing the array around the scala tympani, and then examining the bones histologically after some weeks.

Our studies [32, 34] showed that less trauma occurred with an electrode array passed around the scala tympani of the basal turn of the cochlea, rather than with electrodes inserted through holes drilled into the apical and middle turns. This is illustrated in figures 14 and 15 [34].

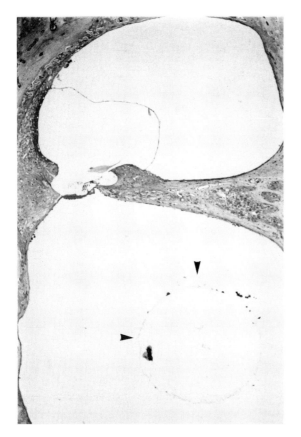

Fig. 14. Photomicrograph of a cat cochlea following the insertion of a round Silastic electrode carrier into the scala tympani of the basal turn through an incision in the round window membrane. There is no visible damage to the cochlea or loss of spiral ganglion cells 18 weeks postimplantation. ×51.5.

Figure 14 is from a recent study and is a 2-μm section using Spurr's epoxy resin for embedding. There is minimal or no trauma to the cochlea and spiral ganglion cells following the insertion of this free-fitting Silastic electrode carrier. Figure 15 is a celloidin section showing the quite extensive damage and tissue reaction that can occur if electrodes are inserted through holes drilled in the bone overlying the apical and middle turns of the cochlea. This damage is primarily due to the difficulty of locating the drill site so that the electrode can be passed into the scala tympani and not into the scala media or scala vestibuli.

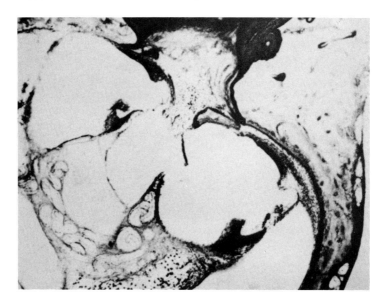

Fig. 15. Photomicrograph of a cat cochlea 58 weeks following the insertion of an electrode through an opening drilled into the middle and apical turns [34]. ×89.

Our studies and those of others have shown that histopathological changes following the insertion of an electrode along the scala tympani of the basal turn of the cochlea can occur in certain circumstances. Firstly, a localized tear of the basilar membrane or fracture of the spiral lamina can result in the loss of spiral ganglion cells and auditory nerve fibres, but this is localized to the site of the lesion [31, 34, 119, 126, 128]. The loss of ganglion cells following a tear in the basilar membrane is shown in figure 16. The tear in figure 16 resulted from a Teflon® (PTFE) strip which was stiff and had sharp edges. This also emphasizes that the mechanical properties of an electrode are important. By way of contrast, figure 17 shows a round Silastic electrode close to the basilar membrane without trauma.

Another important observation was that a localized lesion of the spiral ligament did not lead to any loss of spiral ganglion cells or auditory nerve fibres and this is illustrated in figures 18 and 19. Figure 18 shows a photomicrograph from a cat cochlea where a Silastic tube with a square end caused a lesion in the spiral ligament without the loss of the dendritic

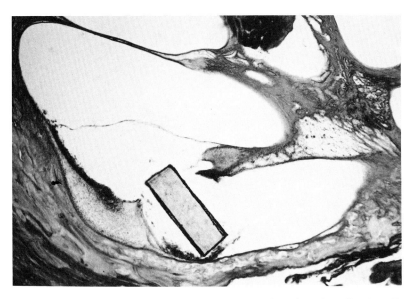

Fig. 16. Photomicrograph of a cat cochlea following the insertion of a Teflon (PTFE) strip into the scala tympani of the basal turn through an incision in the round window membrane. There is a tear of the basilar membrane and loss of spiral ganglion cells 24 weeks postimplantation. ×30.

nerve fibres. A high power view of the dendrites is shown in figure 19. As the cat had been implanted for 6 weeks any significant loss of dendrites would have been apparent, no subsequent loss of spiral ganglion cells after 6 weeks would have occurred as there was no significant loss of dendrites.

Finally calcification could occur within the cochlea following implantation. Calcification was usually confined to a localized area beneath the spiral lamina and was associated with a fibrous tissue inflammatory response. It was also induced when a fracture of the spiral lamina occurred; this is seen in figure 20. In this figure fragments of the fractured spiral lamina can be seen within the area of calcification.

Electrode Insertion in Human Temporal Bones

Although pathological changes in the cochlea can occur in certain circumstances following surgical trauma, they should be largely avoidable. To ensure that this would be the case, we carried out further research to develop an atraumatic surgical approach.

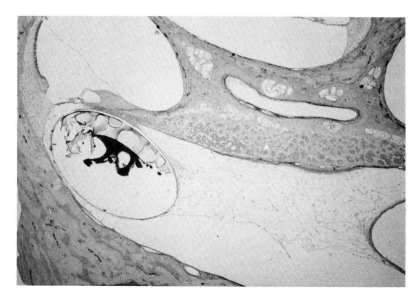

Fig. 17. Photomicrograph of a free-fitting Silastic electrode carrier inserted into the scala tympani of the basal turn of the cat cochlea through an incision in the round window membrane 16 weeks postimplantation. ×45.

In our first study on three control and nine implanted bones we inserted the free-fitting, smooth, tapered, banded electrode array, and then after removing the electrode, we sectioned the temporal bones looking for any microscopic evidence of trauma. In this study [122] localized lesions of the basilar membrane or spiral lamina occurred in three out of nine bones. In each case the lesion was restricted to a small region of approximately 1 mm along the cochlea, indicating that the auditory nerve fibres and ganglion cells would have only been reduced in number over a small proportion of the insertion length. Furthermore, in each case the tear or fracture only occurred when the surgeon continued to apply force after resistance was felt. Consequently, the lesions could be prevented if the surgeon made it a rule not to apply force after resistance was felt.

Another type of lesion seen in some of the temporal bones was a small tear in the surface of the spiral ligament. This occurred when the electrode tip first touched the outer wall before it passed around the basal turn. An example of a tear in the spiral ligament in the human temporal bone is shown in figure 21. As discussed above, our animal studies showed

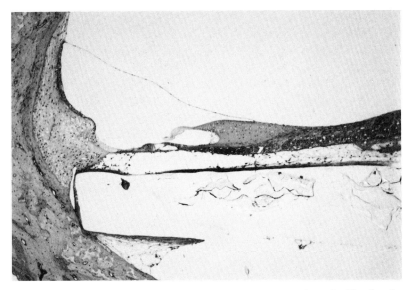

Fig. 18. Photomicrograph of a cat cochlea following the insertion of a Silastic tube into the scala tympani showing a lesion of the spiral ligament but no loss of dendrites 6 weeks postimplantation. ×31.

that a similar lesion did not lead to the loss of spiral ganglion cells or auditory nerve fibres (fig. 18, 19).

Finally, tears were seen in Reissner's membrane in some bones. In order to determine whether they were induced by the histological preparation of the bones or the actual electrode insertion the data were analysed statistically. The results showed there was no significant difference between the control unimplanted bones and those implanted with the electrode array with regard to the number and length of Reissner's membrane tears. Thus the tears were considered to be primarily artefacts.

The findings of the histological study are summarized in figure 22 [122]. These results which were discussed above showed that the multiple-electrode banded array could be inserted safely into the scala tympani of the basal turn of the cochlea with minimal or no trauma providing no force is applied after resistance is felt.

To further examine the safety of electrode insertion two surface preparation studies were carried out where any trauma was observed by removing the overlying bone to show the cochlea and electrode. The first

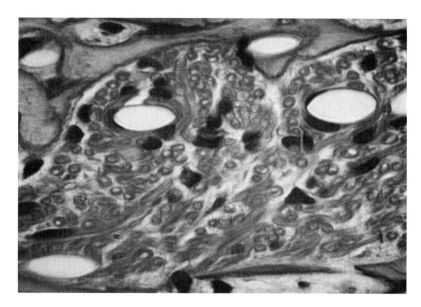

Fig. 19. Photomicrograph of the cat cochlea in figure 18 at a higher magnification confirming the presence of dendrites 6 weeks after a lesion in the spiral ligament. ×12,600.

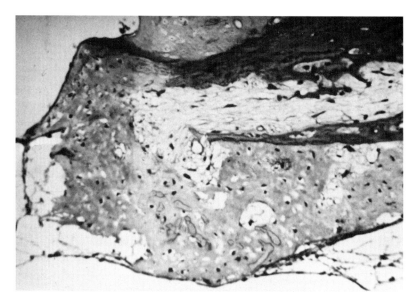

Fig. 20. Photomicrograph of the cat cochlea showing calcification due to a fracture of the spiral lamina. ×360.

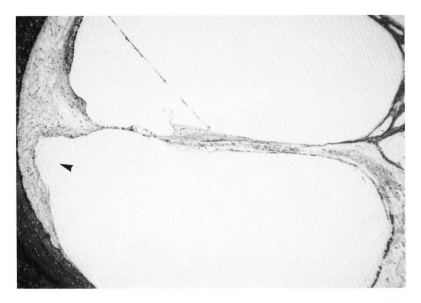

Fig. 21. Photomicrograph of a human temporal bone after the insertion of the Nucleus electrode array showing a tear of the spiral ligament. ×50.

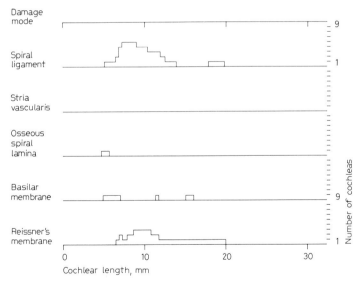

Fig. 22. Summary of the results of the histological study on trauma following electrode insertion in nine human temporal bones. Histograms of the number of cochleas affected by trauma at sites along the basal turn have been plotted [121].

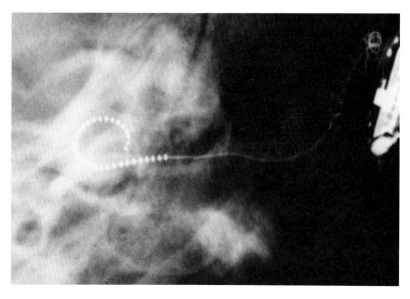

Fig. 23. Plain X-ray of the temporal bone of a cochlear implant patient showing the banded electrode array in the basal turn of the cochlea.

study was at the University of Sydney [54] and the second at the University of Melbourne [52]. These studies were carried out on a total of 24 bones. These surface preparation studies confirmed the findings of the histological study that the electrode can be inserted safely around the basal turn of the cochlea with minimal or no trauma provided no force is applied after resistance is felt. Figure 23 is an X-ray of a patient's cochlea showing how the banded electrode can be inserted around the basal turn of the cochlea in a patient without bending or distorting.

Electrode Mechanical Properties

To ensure that surgical trauma can be kept to a minimum the mechanical properties of the multiple-electrode array have been examined to make sure the tip will flex easily and that the electrode is not too rigid. Furthermore, as trauma has been observed in human cochleas [89] implanted with single platinum wires with a diameter of 0.21 mm [86], the force required to buckle these single electrodes was compared with that required to buckle the Nucleus multiple-electrode array. The measure-

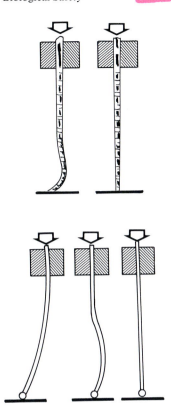

Fig. 24. The measurement set-up to compare the mechanical properties of the Nucleus multi-electrode banded array and a single-channel electrode [86, 110].

ment set-up to compare the buckling stresses and flexibilities of the two electrodes is shown in figure 24. The study was carried out by the Commonwealth Scientific Industrial Research Organisation of Australia and showed that the maximum force applied by the single electrode before buckling was 25 times greater than that by the Nucleus multiple electrode array, and that the Nucleus electrode array was ten times more flexible [110]. This is due to the fact that the Nucleus array is made of fine platinum wires with a diameter of 0.025 mm compared to the single electrode which had a diameter of 0.21 mm. The graded stiffness with 22 wires at the proximal end and only one near the tip (fig. 25) results in more flexibility. Consequently, the histopathological findings reported by Johnsson et

Fig. 25. Photograph of the tip of the Nucleus multi-electrode banded array showing the fine platinum wires being added sequentially [51]. ×21.5.

al. [89] apply to a reasonably rigid electrode system, but not the Nucleus multiple-electrode array.

Further Anatomical Studies

Additional anatomical studies have been undertaken to further minimize surgical trauma and to maximize the depth of the electrode insertion before resistance is felt. The first of these studies was a detailed examination of the anatomy of the basal turn of the cochlea near the round window. This showed that in order to get a good view of the basal turn and in order to pass the electrode without it coiling on itself, either the ridge of bone at the round window (the crista fenestrae – fig. 26) needs to be drilled away, or a separate opening made into the basal turn just antero-inferior to the round window [79]. If the crista fenestrae is drilled away the electrode may pass more readily around the basal turn of the cochlea (fig. 27). If this is not done the electrode can meet the outer wall at an acute angle, in which case it is likely to bend on itself or be deflected towards the basilar membrane.

Another study [78] was undertaken to examine in detail the motion of the electrode as it passed around the scala tympani of the basal turn of the cochlea. As a result of this work we found that if the electrode is rotated through 90–180° (anticlockwise for the right and clockwise for the left) after it has been inserted for 10 mm, the tip will be directed away from the basilar membrane and it will more easily pass around the basal turn of the cochlea. This is illustrated in figure 28.

Explantation and Reinsertion

We have also been concerned to know whether the smooth, free-fitting, banded, electrode array can be easily withdrawn and another rein-

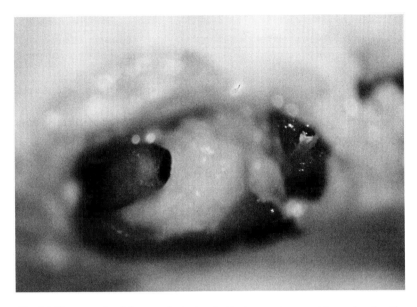

Fig. 26. Photograph of the round window in the human temporal bone showing the crista fenestrae at the antero-inferior margin of the window [79].

serted with minimal trauma. This is important information as failure modes, although not likely, may be distal to the electronics package, in which case the electrode would need to be replaced. Secondly, eliminating a connector would help keep the package thin for use in children.

We have carried out explantation and reinsertion studies in animals and found that the electrode comes out easily and another one can be reinserted without difficulty. We have also been able to study the explantation and reinsertion of the banded electrode array in three patients who had the banded electrode array and the University of Melbourne's prototype receiver-stimulator package replaced with a similar banded electrode array connected to the Nucleus receiver-stimulator [52]. In these patients the electrode array came out easily, and another array could be inserted for at least the same distance. Furthermore, the patients' clinical results were not adversely affected by the explantation and electrode reinsertion. These results imply that it is feasible to explant and reinsert an electrode array without compromising the clinical effectiveness of the cochlear implant.

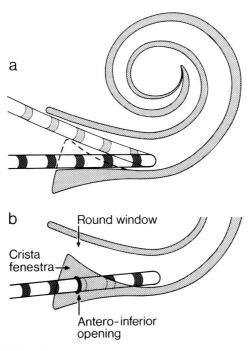

Fig. 27. Diagrams showing the direction of electrode insertion through the round window and along the basal turn both with and without the crista fenestrae drilled away [79].

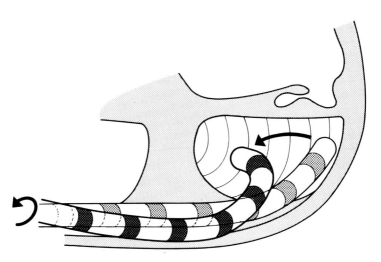

Fig. 28. Diagram showing how rotation of the electrode directs the tip away from the basilar membrane and around the basal turn of the cochlea [78].

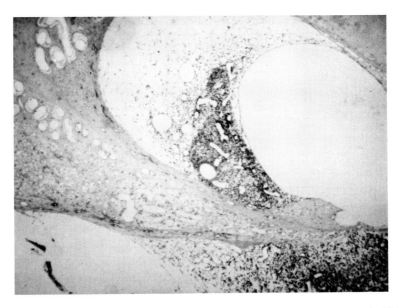

Fig. 29. Photomicrograph of the cat cochlea showing an acute infective labyrinthitis postimplantation. ×50.

Summary

In summary, the studies on an atraumatic surgical approach have shown that the smooth, free-fitting, banded, electrode array can be inserted along the scala tympani with minimal or no trauma providing no force is applied when resistance is felt. Learning the force that can be safely applied to the electrode is best practised in the temporal bone laboratory.

The Prevention of Infective Labyrinthitis Postimplantation

Our initial studies [32, 34] showed that infection in the cochlea could occur following implantation (fig. 29), and this could lead to a severe loss of spiral ganglion cells. Consequently, we considered it important to carry out studies to determine the spread of infection to the implanted cochlea and its prevention, and also to develop a surgical protocol that would minimize the risk of a postoperative infection. These studies are listed in table 7.

Table 7. Studies on infective labyrinthitis postimplantation: aetiology and prevention

1   Animal model of otitis media
2   Spread of infection to the cochlea
3   Barriers to infection
4   Surgical protocol for prevention of infection

Animal Model of Otitis media

Studies on infection and its prevention have been undertaken firstly to develop an animal model of otitis media, secondly to investigate the spread of infection to the cochlea, and thirdly to evaluate potential barriers to the spread of infection. We have been developing animal models of otitis media so that we could study the effects of the organisms that most frequently cause otitis media in man. We consider it important to study representative organisms because they have different pathological characteristics. For example, *Streptococcus pneumoniae* and *Haemophilus influenzae* organisms are encapsulated which makes them resistant to phagocytosis, *Streptococcus pyogenes* produces lytic enzymes which can facilitate its spread, while *Staphylococcus aureus* on the other hand produces a coagulase enzyme which tends to wall off the infection and prevent access from the body's defence mechanisms.

Spread of Infection to the Cochlea

Having developed a model in the cat for producing a severe otitis media with *Staphylococcus aureus, Streptococcus pyogenes* and more recently *Streptococcus pneumoniae* organisms [4, 5, 19, 77], we have been studying the conditions under which otitis media would lead to infection in an implanted cochlea, and to what extent the electrode sheath and round window would act as a barrier.

We have implanted cochleas, induced otitis media and examined the cochleas histologically for the spread of infection. The results have shown that the healed round window membrane and the fibrous tissue sheath around the electrode could resist the spread of infection to the cochlea especially for *Staphylococcus aureus* and *Streptococcus pyogenes* organisms [19, 77]. We have as yet insufficient data to be certain about *Streptococcus pneumoniae* [4, 5]. The protection offered by the round window membrane to *Streptococcus pyogenes* is shown in figure 30.

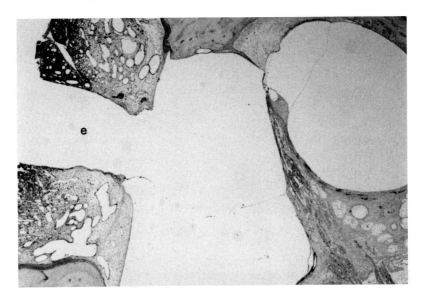

Fig. 30. Photomicrograph of the implanted cat cochlea showing the round window and basal turn 14 days following inoculation of the bulla with *Streptococcus pyogenes* organisms. The acute inflammatory response in the bulla can be compared with the near normal condition of the scala tympani and highlights the effectiveness of the round window seal [121]. e = electrode. ×50.

Barriers to Infection

Although an otitis media induced by *Staphylococcus aureus* and *Streptococcus pyogenes* infection postimplantation does not usually result in a labyrinthitis, it may do so on occasions. For this reason we have been studying the factors that can predispose to infection extending from the middle to inner ears and their prevention. This has involved examining seals produced around the electrode entry point over a period of time, as well as its response to infection [57]. We have compared the seals produced around an electrode inserted through an opening in the round window membrane or the bone just antero-inferior to the round window [77]. We have also investigated seals resulting from heterologous material glued around the electrode shaft to form a plug at the electrode entry point [49, 76]. The only consistent finding in these studies has been that a labyrinthitis is most likely to develop in the first few weeks after implantation before a seal has adequately formed.

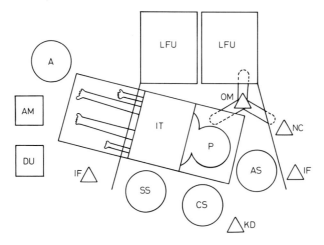

Fig. 31. Diagram of operative layout for a multiple channel cochlear implant operation [38]. A = Anaesthetist; AM = anaesthetic machine; AS = assistant surgeon; CS = chief surgeon; DU = diathermy unit; IF = irrigation fluid; IT = instrument table; KD = drill; LFU = laminar flow unit; NC = nitrogen cylinder; OM = operating microscope; P = patient; SS = scrub sister.

Surgical Protocol for the Prevention of Infection

In view of the risk of an infection in the early postoperative phase, we have developed a surgical protocol to minimize the risk of infection and its spread to the cochlea in the first 2 weeks postimplantation. As part of this protocol we operate in a horizontal laminar flow of filtered air to reduce the risk of bacteria falling into the wound. Our studies [38] have shown that this is only effective if the operating table, drapes, and instruments are correctly positioned as shown in figure 31. In addition, not only is there strict asepsis, but we also infiltrate the operative field with a dilute solution of ampicillin and cloxacillin and administer these drugs parenterally.

Safety of Electrical Stimulus Parameters

Finally, a very important part of our safety studies has been to ensure that the electrical stimulus parameters used do not lead to damage of the spiral ganglion cells. These studies are summarized in table 8.

Table 8. Safety of electrical stimulus parameters

---

1 In vitro studies
   A Platinum dissolution
      Spectrophotometry
   B Electrode corrosion
      Scanning electron microscopy
2 In vivo studies
   A Electrode corrosion
      Scanning electron microscopy
   B Ganglion cell viability
      Electrically evoked brainstem response
      Light microscopy
      Transmission electron microscopy

Damage to spiral ganglion cells may occur if toxic or corrosion products are produced at the electrode, or if the electrical stimulus parameters themselves have a direct effect on the ganglion cells.

In vitro Studies
Platinum Dissolution
Initially we carried out in vitro studies to help establish the electrical stimulus parameters that would lead to minimal corrosion of platinum electrodes. Corrosion of platinum was measured by a spectrophotometric technique. This technique depends on the fact that stannous chloride forms a yellow-to-orange soluble product with platinum ions, and that the absorption of light using a spectrophotometer is proportional to the platinum ion concentration. The results are shown in figure 32 [8]. In this figure total dissolved platinum versus pulse duration is shown for different current densities. The dissolved platinum increased with both current density and pulse duration. It is important, therefore, to keep current density low, and pulse duration short. For this reason, we use a banded electrode, as it is circumferential and has a large area, and consequently the current density is kept as low as possible. A banded electrode is shown in figure 33. For these studies we also use a pulse width that is normally 180 μs/phase. In addition, other studies [24] have shown that it is important to have a biphasic stimulus pulse, and to keep the charge balanced between the two phases.

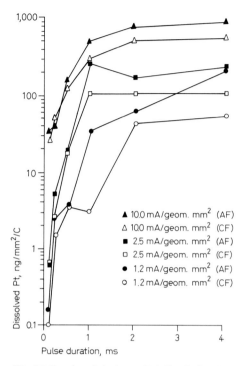

Fig. 32. Results of platinum (Pt) dissolution for biphasic current pulse stimulation at current densities of 1.25, 2.5 and 10.0 mA/geom. mm^2 as a function of pulse width per phase [8].

Electrode Corrosion

We have examined the banded electrode under the scanning electron microscope following in vitro stimulation in normal saline, using stimulus parameters that minimize corrosion [123]. We have found, however, that pitting occurred. This is shown in figure 34. This occurred following stimulation at a charge density of 0.36 μC mm^{-2} geom. per phase for 500 h.

In vivo Studies
Electrode Corrosion

Fortunately, in contrast to the in vitro situation, no significant corrosion occurred following in vivo stimulation at charge densities within the

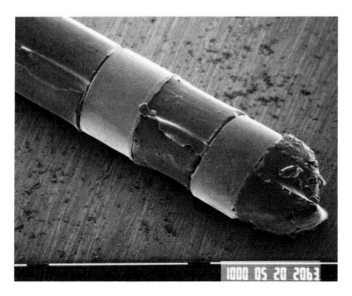

Fig. 33. Scanning electron micrograph of the banded electrode. Bar = 1,000 µm.

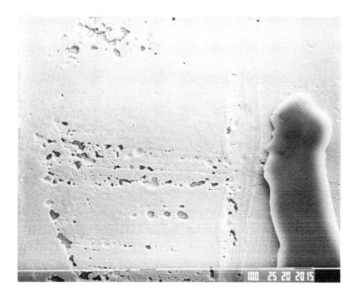

Fig. 34. Surface of in vitro electrode following 500 h of stimulation in inorganic saline at a charge density of 0.36 µC/mm^{-2} geom. per phase. Extensive surface pitting in the form of corrosion is apparent [123]. Pt = Platinum electrode; S = Silastic; bar = 100 µm.

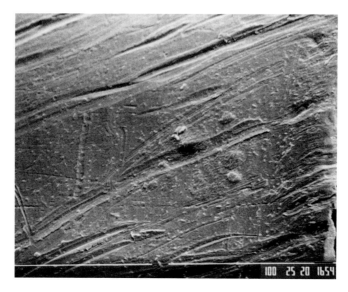

Fig. 35. Surface of in vivo electrode (108R) following 1,189 h of stimulation, at a charge density of 0.32 μC mm^{-2} geom. per phase. Although fabrication marks were apparent, pitting corrosion was not present [122]. Bar = 100 μm.

operating range of the cochlear prosthesis. Figure 35 is a scanning electron micrograph of an electrode following in vivo stimulation at a charge density of 0.32 μC mm^{-2} geom. per phase. The lack of corrosion with in vitro stimulation confirms the findings that protein can inhibit the platinum corrosion during periods of chronic stimulation.

It is also interesting that no corrosion occurred in spite of the surface excoriations due to the fabrication of the platinum sheet or the working of the material during its incorporation as a banded electrode in the array. Normally these excoriations will facilitate corrosion at points where charge density is concentrated. Nevertheless, to avoid surface excoriations facilitating corrosion, the manufacturing technique for the Nucleus electrode array was changed and the electrodes are now made from platinum tubing which does not have this problem. This technique also reduces the extent of cold metal working which predisposes to metal fatigue. A scanning electron micrograph of the electrode surface of the present Nucleus array in figure 36 shows very little excoriation.

Fig. 36. Scanning electron micrograph of a platinum (Pt) electrode manufactured from Pt tubing. Electrodes manufactured this way show significantly less mechanically induced damage compared with electrodes manufactured from Pt foil [122]. S = Silastic; bar = 100 μm.

Ganglion Cell Viability

Having determined the electrical stimulus parameters that would minimize electrode corrosion, we carried out in vivo studies on cats to ensure that the stimulus parameters used would not lead to loss of ganglion cells. Cats were stimulated continuously for periods of up to 2,029 h at stimulus levels half way between threshold and discomfort level and at a rate of 500 pulses/s. During the experiment electrically evoked EABR, were recorded so we could correlate any changes with subsequent histology of the cochlea [120].

Figure 37 shows the EABR for a typical cat, No. 99, that had 1,348 h of stimulation at a current of 0.8 mA and charge density of 0.28 μC mm^2 per phase. In this figure current is plotted against EABR amplitude for different stimulus durations. The input/output functions have two limbs. The limb with the gentle gradient is due to electrophonic stimulation of hair cells, and the limb with the steep gradient is due to direct stimulation of auditory nerve fibres [12]. Notice that the input/output functions

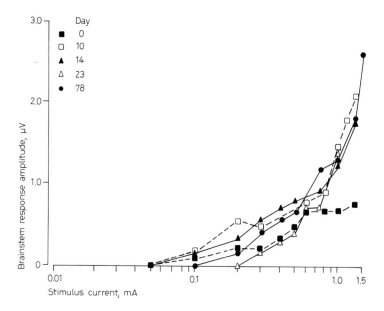

Fig. 37. Electrically evoked auditory brainstem responses plotted against stimulus currents for different stimulus durations up to 1,348 h for cat 99 [120].

remain stable over 1,348 h (78 days) of stimulation. There appears to be no loss of hair cell function shown by the preservation of the elec- trophonic limb, and there also appears to be no loss of spiral ganglion cells shown by the stability of the steep gradient limb of the input-output function over the duration of the stimulus period. These EABR findings were confirmed by the histology. Figure 38 is a low power view of the cochlea in the stimulated area showing the fibrous tissue capsule of the electrode. The organ of Corti was found to be normal throughout all turns of the cochlea, and the spiral ganglion cells were also intact. Figure 39 is a high-power view of the spiral ganglion cells showing them to be normal.

When we plotted the mean spiral ganglion cell densities against stimulus duration we found no statistical correlation (figure 40). There was also no correlation between the stimulated and control sides. These findings, therefore, showed that the stimulus parameters used did not lead to a loss of spiral ganglion cells, and this was consistent with the EABR results. The only correlation found was between mean spiral ganglion cell

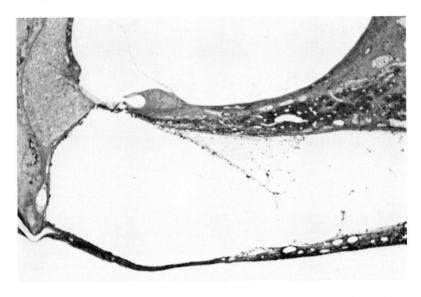

Fig. 38. Light micrograph of the cochlea of cat 99 from the stimulated area [120].
×26.

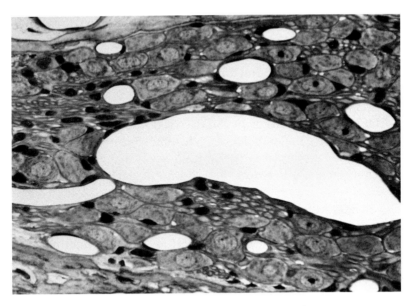

Fig. 39. Light micrograph of spiral ganglion cells of cat 99 from the stimulated area
[120]. ×300.

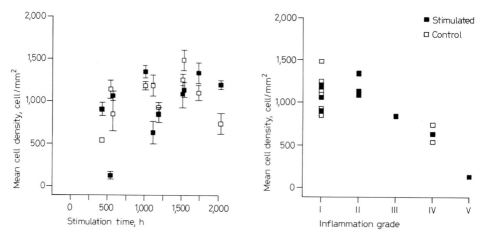

Fig. 40. Mean spiral ganglion cell densities plotted against stimulation time and inflammation grade. Inflammation was graded on the basis of the number of polymorphonuclear and mononuclear leucocytes, and the thickness and maturity of the fibrous tissue [120].

density and the severity of an inflammatory reaction due to infection. This confirmed the previous findings that infection causes a marked ganglion cell loss. Furthermore, the incidence of infection was the same for both stimulated and control bones, which showed that electrical stimulation per se did not predispose to a labyrinthitis.

Another important finding in this study was that calcification in the cochlea was not produced by the electrical stimulus parameters used. It was seen in an equal number of both stimulated and unstimulated control bones. Furthermore, calcification was usually limited to an area beneath the spiral ligament, and did not appear to affect the EABR thresholds.

In addition to our in vivo safety studies on current density we have more recently been examining the effect of rate of stimulation. This is a particularly important parameter to study as some multiple-electrode implants using band-pass filtering techniques and most single electrode systems stimulate at quite high rates.

We carried out the study on cats by recording the EABR amplitudes to different probe currents, presented following periods of electrical stimulation. The probe currents were 0.5, 1.0, and 1.5 mA. The EABR

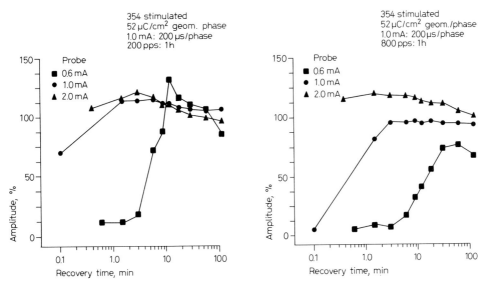

Fig. 41. EABR amplitudes poststimulation for probe currents of 0.5, 1.0 and 1.5 mA following stimulation at rates of 200 and 800 pulses/s.

for the smallest probe current were depressed to a greater degree than for the larger probe currents, as the smallest current tested the excitability of the nerve fibres in closest proximity to the stimulating electrodes. The results following stimulus rates of 200 and 800 pulses/s are shown in figure 41. At 200 pulses/s there was a temporary stimulus-induced change in the EABR for the smallest monitor current, and this returned to normal within 10 min. On the other hand results for stimulation at a rate of 800 pulses/s show there was a permanent stimulus-induced change in the EABR for the smallest monitor current.

These findings have implications for speech processors which produce high rates of stimulation to code speech frequencies and indicate that more safety data is required for cochlear implants stimulating at high rates. This is one reason the Nucleus speech processor only codes low frequency voicing as rate of stimulation up to 200 pulses/s, while higher formants are coded as place of stimulation without increasing the rate of stimulation. Furthermore, the Nucleus speech processor does not stimulate continuously at one site as is the case with other speech processors. Our experimental studies have also shown that periods of rest also reduce the permanent effects of stimulation.

Summary

In summary, the studies on the biocompatibility of materials, an atraumatic surgical approach, the prevention of infective labyrinthitis postimplantation, and electrical stimulus parameters, show that intracochlear multiple-electrode stimulation is safe providing care is taken with the procedure as described in the above sections.

The Engineering of the Receiver-Stimulator and Speech Processor

The engineering development of the receiver-stimulator and speech processor are summarized in table 9.

Receiver-Stimulator

The physiological, psychophysical and histopathological studies on the experimental animal discussed above provided data for the design [75] of the University of Melbourne's prototype receiver-stimulator first implanted on 1st August 1978.

The information showed, firstly, that as there were limitations in coding pitch by rate or timing of stimulation it would be necessary to code pitch on a place basis as well, in order to maximize speech understanding. To do this it would require implanting a number of electrodes in the cochlea so that discrete groups of auditory nerve fibres could be stimulated. This would require either transmitting speech information to the multiple-electrode arrays by a direct link through the skin (percutaneous plug) or by implanting a receiver-stimulator unit and sending the information transcutaneously through the intact skin.

Transcutaneous vs Percutaneous Link
It was decided in 1972 to use transcutaneous stimulation for a number of reasons. Firstly, our experience with the long-term use of percutaneous plugs in animals had shown that there was a tendency for the skin to become macerated at its junction with the plug, and also to produce a sinus in the skin. These provided a nidus for infection which could track around the foreign body and become chronic. Eradication of the infection became difficult and it had to be controlled with the topical application of antibiotics or the removal of the foreign material. Secondly, the percutaneous plugs could be easily damaged and the electrodes dislodged or fractured. Thirdly, it was considered that percutaneous plugs would not be aesthetically acceptable, especially in children.

Table 9. The engineering of the receiver-stimulator and speech processor

---

1 Receiver-stimulator
   A Transcutaneous vs percutaneous link
   B Type of transcutaneous link
   C Analogue vs digital circuits
   D Data and power transfer
   E Electronic design
   F Packaging
   G Connector
   H Lead wire assembly
   I Biocompatibility
   J Reliability
   K Smaller version for children
2 Speech processor
   A Laboratory based speech processor
   B Portable prototype speech processor
   C Wearable speech processor for clinical trial
   D Diagnostic programming system

Transcutaneous Link

The next decision was how to transmit the coded speech information through intact skin to an implanted receiver-stimulator. Should this be carried out by an electromagnetic link, by optical signals, or ultrasonically? After consideration it was decided to use electromagnetic induction as the most efficient and reliable method. Electromagnetic induction is based on the principle illustrated in figure 42 that a magnetic flux produced by passing a current through a coil (external) will induce a current in a second coil (internal).

Adequate power must be transferred from the external to internal coils over a distance that will vary depending on the thickness of patient's skin and underlying tissues. It is desirable that some lateral misalignment be possible between the two coils. Even if magnets in the centre of both coils are used to help the patients bring the external coil into the correct position, precise coaxial alignment will be difficult, or the coil may be displaced. There should be adequate power transfer over a distance of up to 10 mm when the coils are coaxial, with some degree of misalignment possible at a shorter distance.

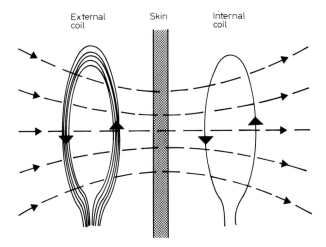

Fig. 42. Diagram of an external transmitter coil with electromagnetic energy passing through the skin to induce a current in the internal receiver coil [50].

Analogue vs Digital Circuits

In transmitting coded speech signals and power another important decision was whether to use analogue or digital circuitry or a combination of both in the design of the receiver-stimulator unit.

Analogue circuits are those where continuously varying physical parameters such as voltages can be altered or combined. With analogue circuitry the instantaneous amplitude of speech could be converted into a voltage proportional to the amplitude. The voltage could then be transmitted to the receiver-stimulator where the induced voltage (or current) would be used to stimulate nerve fibres. On the other hand, with digital circuitry the frequency and instantaneous amplitude of speech would be transmitted and extracted as a digital code. The digital code would be represented by discrete pulses, and could be quantified as a series of digits or numbers which could then be manipulated mathematically.

There were and still are a number of advantages in using a digital system. Firstly, it is straightforward to combine the control information for each electrode pair into a single signal, and to recover this information in the receiver-stimulator. A single transmission path or pair of induction

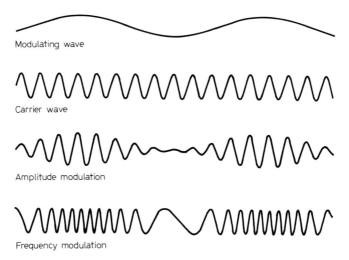

Modulating wave

Carrier wave

Amplitude modulation

Frequency modulation

Fig. 43. Diagram of amplitude and frequency modulated carrier waves [50].

coils can then be used for a multiple-electrode implant. On the other hand, with a multiple-electrode analogue system, separate coils would be needed for each electrode pair in order to minimize power consumption. This would mean a bulky receiver-stimulator system which would be awkward to implant, especially in children. Secondly, digitally controlled current sources are more reliable with high noise immunity and would therefore deliver well-defined stimuli. The speech processor could be precisely adjusted to suit individual patients.

In view of the above advantages and the ready availability of digital designs in integrated circuit silicon chip technology a digital receiver-stimulator system was realized.

Data and Power Transfer

Data and power were transmitted by modulating a carrier wave. This is illustrated in figure 43 where examples of amplitude and frequency modulation are shown. The modulated wave was decoded by the implanted receiver-stimulator electronics, and the power and coded speech signals extracted. In the case of the prototype receiver-stimulator power was transmitted at a carrier frequency of 112 kHz and speech data at 10.752 MHz [33, 35]. With the device manufactured by Coch-

Fig. 44. Photograph of the prototype receiver-stimulator with connector developed by the University of Melbourne and first implanted on 1st August 1978. ×1.7.

lear Pty Limited power and coded speech data are transmitted on a carrier frequency of 2.5 MHz. The implantable prototype receiver-stimulator used in our first patient in 1978 is shown in figure 44, and the device developed by Cochlear Pty Limited for clinical trialling is shown in figure 45.

Electronic Design

The design of the prototype receiver-stimulator [33, 35] needed to allow for as much flexibility as possible in the use of stimulus parameters as we were not sure how to produce a speech-processing strategy that would enable patients to understand running speech, or the range of parameters required. It was designed so that intensity could be varied from a minimum of 70 µA to a maximum of approximately 1 mA in 70-µA steps. The rate of stimulation could be varied in 125-µs steps up to 1 kHz on each of ten stimulus channels, and the phase relations between pulses could be varied in eight 125-µs periods [33, 35].

The design of the prototype receiver-stimulator was realized using a hybrid circuit in which various commercially available components as well as a custom-made integrated circuits were interconnected on three substrates using thick film technology (fig. 46).

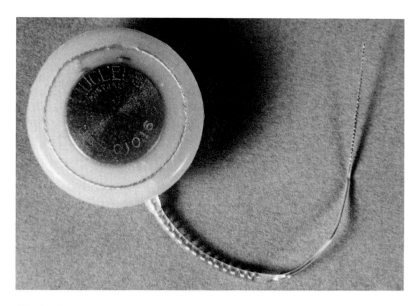

Fig. 45. Photograph of the receiver-stimulator for clinical trial developed by Cochlear Pty Limited in conjunction with the University of Melbourne and first implanted on 13th September 1982. ×1.4.

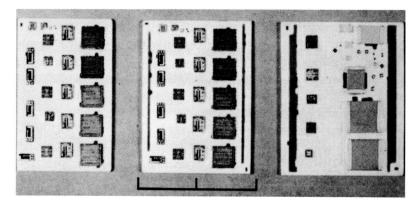

Fig. 46. The hybrid circuits for the University of Melbourne prototype receiver-stimulator. ×1.7.

As a result of the psychophysical and speech perception studies on the first two patients it was discovered that more control of intensity was desirable, and as variations in loudness with simultaneous stimulation on two or more stimulus channels could not be adequately predicted without more research, non-simultaneous bipolar stimulus channels were needed. The number of frequency steps and upper limit on rate of stimulation were adequate.

To provide more amplitude information the Nucleus receiver-stimulator, manufactured for clinical trial, was designed so that the stimulus current pulse could be adjusted in amplitude over the range from 25 µA to 1.5 mA in 3% steps. The wide range of currents was selected to accommodate anticipated differences in patient thresholds and discomfort levels due to variation in electrode position, population of residual nerve fibres, and the effect of fibrous tissue or bone growth. The 3% incremental steps in amplitude were of the same order as the reported amplitude difference limens.

The prosthesis has a facility whereby the pulse width can be varied, and this is a useful addition as this stimulus parameter can be used as well as current level to produce changes in loudness. It has, in fact, been shown that it is possible to trade off current level for pulse width and maintain equal loudness providing total charge is constant. With this receiver-stimulator the pulse width could be varied from 20 to 400 µs per phase in steps of 0.4 µs.

As it was difficult to predict loudness following simultaneous stimulation, the receiver-stimulator for clinical trial was designed to provide non-simultaneous stimulation at each electrode pair. This means that no stimuli at different pairs of electrodes overlap in time. There is always a short time interval between the stimuli so there is no interaction between electrical fields producing unpredictable variations in loudness. Furthermore, as simultaneous stimulation was not used the need for the control of phase between stimuli was not required.

The implant for clinical trial was designed to provide bipolar stimulation on 22 electrode pairs as well as common ground stimulation. Furthermore, with bipolar stimulation the pairs could be either adjacent electrodes or alternatively the current could be made to pass between electrodes separated by one or more intervening electrodes.

Finally, the receiver-stimulator was designed to allow stimulus rates in excess of 1,000 pulses/s and was implemented on a single silicon chip using digital and analogue circuitry.

Packaging

The packaging of the receiver-stimulator electronics was considered to be very important, and this was carried out using an hermetically sealed container impervious to body fluids. The standards needed to be high as body fluids and enzymes permeate along minute pathways or cracks. The sealing needed to be validated using helium leak testing. Helium was used as it can be detected in very small concentrations, and has a fast diffusion rate.

A great deal has been learnt from pacemaker technology about sealing electronic units, and this was applied to the Nucleus implant. For example, epoxy resins were originally used for pacemakers, but led to later electronic failures. Kovar containers with glass feed-throughs and sealed by soldering have been used in space programmes. They were also used for packaging our first prototype multiple-channel implants. Although these containers were satisfactory for space flights, the body is a more hostile environment. It is quite likely to erode the glass insulation around the wires entering and leaving the package, or surface tension forces may enlarge small cracks in the glass so they become fluid entry paths. Furthermore, with time, the metals in the solder can migrate or produce corrosion from an electrolytic reaction and thus lead to weaknesses in the seal.

The most reliable packaging material is titanium which has now been used for many years in heart pacemakers. It is an inert metal and the seal can be made by welding the edges of the two halves of the containers together. In this way the seal is made of the same material as the container so no electrochemical reaction can be set up, and cause corrosion.

The use of titanium or other metals, however, presents two problems. Firstly, there needs to be an effective seal and adequate insulation at the points where the wires enter or leave the package. Glass is not really effective for long-term use in the body and ceramics are to be preferred. Ceramics have been found reliable in heart pacemakers, and can form a good bond with metal. In the Nucleus receiver-stimulator for clinical trial the electronics were packaged in a titanium capsule and special ceramic feed-throughs have been developed. Unfortunately the receiving coil cannot be placed inside the package as electromagnetic energy cannot be transmitted through the metal. The coil was, however, placed around the titanium capsule, as illustrated in figure 45, without actually increasing its thickness.

The receiver-stimulator should be designed for ease of surgical placement and cosmetic acceptability. Surgical experience as a result of implanting our first prototype device, which was rectangular in shape, as well as a series of more than 50 implants since 1982 using the Nucleus receiver-stimulator for clinical trial, which is round, has confirmed the view that a round device is to be preferred. The bed can be made very neatly with a milling burr [48]. A round shape is also desirable for other reasons. It conforms best to a circular receiver coil, and it has no sharp angles where stress concentrations can occur as a result of welding the two halves together leading to inadequate sealing.

Our anatomic and surgical studies showed the maximum depth a bed could be drilled in the mastoid and parietal and/or occipital bones was 6 mm, 3–4 mm was more usual. The maximum height superficial to the bone that is cosmetically acceptable is about 5–6 mm. Acceptability could be improved by rounding off the edges around the package. It was also found that the maximum diameter of the device that could be comfortably placed in adults was about 35–40 mm. With the clinical trial package shown in figure 45, stability was helped by making the implant mushroom-shaped. It has the following dimensions: the titanium capsule and connector have a diameter of 20 mm making up the stalk; the receiving coil has a diameter of approximately 30 mm helping to make up the cap.

Connector

The receiver-stimulator may need a connector so that it can be replaced if there is an electronic failure. It should be added, however, that with heart pacemakers the least likely failure is an electronic one and the same situation will most probably occur with cochlear implants. Connectors therefore are not likely to be essential, especially as reinsertions are possible with minimal damage to cochlear tissues and auditory neurones [52, 53].

If a connector is used pressure between contact points needs to be maintained for many years, and the same metals need to be used to avoid corrosion. Furthermore, the connector should be designed so that body fluids cannot enter and cause current leakage between the electrodes.

The connector used for the prototype receiver-stimulator (fig. 44) consisted of a pair of substrates with conductor patterns printed on them, and layers of conducting and non-conducting elastomate material between the two substrates. The elastomate was compressed between the substrates thus connecting matching sections of the conductor patterns.

It was found that the elastomate lost compression over time. The receiver-stimulator for clinical trial, however, had an improved connector which had an intermediary pad locked into place between the package and electrode assembly pad. The pad was designed to reduce the possibility of loss of compression and electrode contact over time.

Lead Wire Assembly

The packaging of the receiver-stimulator should also be carried out so that it is mechanically robust. Our first series of implants in 1978–1979 showed that the area most vulnerable to repeated small body movements was the point where the electrode array emerged from the package. Any movements transmitted from rubbing the skin or adjusting the transmitter coil would result in maximum bending at this junctional area. Consequently, the Cochlear Pty Limited implant was designed to incorporate stress relief of the electrode wires emerging from the receiver-stimulator so that metal fatigue and fractures of electrode wires would not develop months or years after implantation. This was achieved by tapering the bundle after it emerged from the package, and spiralling the electrode wires. This was incorporated in the receiver-stimulator for clinical trial. It is also recommended [48] during implantation, that this bundle of electrode wires be placed in a groove under the mastoid cortex, and to fix the proximal lead wire assembly to prevent movements from outside being transmitted to the thin distal electrode wires.

Biocompatibility

The receiver-stimulator for clinical trial was moulded into its mushroom shape using Silastic MDX-4-4515 which was shown to be biocompatible. The biocompatibility of the Silastic MDX-4-4515 was similar to that of Silastic MDX-4-4210 which was used as the carrier for the electrode array.

When units were assembled according to the good manufacturing processes laid down by the US FDA, five were implanted intramuscularly in cats for periods of approximately 5 weeks to assess their biocompatibility as discussed under the section on biological safety.

Reliability

The assembled receiver-stimulator units for clinical trial were subjected to a number of environmental tests to ensure their reliability under as many conditions as possible. Electric and magnetic field susceptibility testing was conducted in accordance with a voluntary standard developed by the FDA entitled *Electromagnetic Compatibility Standard for Medical Devices* [146].

The results showed that, in an electric field in the frequency range 10 kHz to 1,000 MHz at field strengths up to 10 V/M, the prosthesis operated normally, except that a low level signal equivalent to a background acoustic signal could be perceived. In a 60-Hz magnetic field having a flux density of 1 mT the prosthesis operated normally, except that a low-level signal equivalent to a background acoustic signal could be perceived. The cochlear implant had more than 100 dB of immunity against the maximum 60-Hz field levels recommended in the Electromagnetic Compatibility (EMC) guidelines. It also had more than 36 dB immunity against the maximum field strengths beyond 5 MHz as recommended in the EMC guidelines.

This immunity to electromagnetic fields can be seen in practice if we consider that a hair dryer has a field in the range 10–20 Gs at 60 Hz, and that the implant is immune to more than 10,000 Gs at this frequency.

The other environmental tests carried out were for dry heat, cold, thermal cycling, free fall, impact shock and vibration. With these tests the performance of the electronic circuit was evaluated before and after the procedure, and X-ray and hermeticity checks were made to ensure there were no adverse mechanical effects. With dry heat the units were evaluated according to AS 1099 test B [147]. The test specimens were introduced to a chamber at ambient conditions and the temperature raised to $100 \pm 2°C$ over a 2-hour period. The temperature was maintained at $100 \pm 2°C$ for another 2 h. The temperature was reduced to $53.9°C$ and maintained for 12 h. The relative humidity was controlled to below 10% and the chamber was then returned to ambient conditions. With the cold assessment, the test was performed according to AS 1099 test A(a) [147]. The test specimens were introduced into the chamber at ambient temperature. During a 1.5-hour period the chamber temperature was reduced to $-20 \pm 2°C$ and maintained at this level for 4 h. The temperature was then raised to $5 \pm 2°C$ over a 1-hour period and kept at this temperature for 12 h. At the 11.5-hour mark performance checks were made. The temperature was then allowed to return to ambient conditions and tested after a 1-hour stabilization period. With termal cycling the test was performed according to MIL-STD-202E method 102A (condition B). The units were subjected to five cycles of alternation from $50°C$ to $-40°C$ with 30 min at each temperature and 15 min between. With free fall the test was performed according to AS 1099 test E(d). Each test specimen was dropped once from a height of 1 m onto each of its six faces. With impact shock the test was performed according to MIL-STD-202E method 213B (condition E). The units were subjected to 18 shocks of 0.6 ms half-side pulses at 50 gravity units, with three shocks on each of six faces. Finally, with vibration the units were vibrated from 5 to 50 Hz at an amplitude of 4 mm dA (5–25 Hz) and 5 gravity units (25–500 Hz) with a sweep time of 30 min over three axes with one sweep per axis.

In all these tests the three receiver-stimulator units evaluated performed normally and there was no evidence of malfunction or adverse mechanical effects from these tests.

Additional tests were also carried out to ensure the durability of the receiver-stimulator for clinical trial. The welds between the electrode bands and the wires in the carrier were subjected to a detailed metallurgical assessment which confirmed the integrity and strength of the welds.

The electrode lead wire was subjected to a stretch and flex test in which the electrode was vigorously flexed more than ten million times, and stretched by more than 50% of its original length. This did not lead to breakages of the lead wires.

Finally a study was carried out on possible failure modes and this has shown that there were no failures that could lead to a life-theatening or hazardous situation.

Smaller Version for Children

Not only has the Cochlear Pty Limited receiver-stimulator discussed above been developed for use on adult patients, but a smaller 'mini' version with magnet for implantation in children has recently been fabricated (fig. 47). The receiver-stimulator discussed above, and shown in figure 45, is satisfactory for patients down to the age of 10 years. It is, however, necessary for children under 10 years to have a smaller receiver-stimulator as well as a magnet incorporated into the device so that a headset with magnet can be easily positioned on the surface of the scalp overlying the implant.

In the 'mini' receiver-stimulator shown in figure 47, the electronics for stimulating 22 electrodes are enclosed in a hermetically sealed titanium capsule similar to that used for the adult receiver-stimulator. The 22 platinum lead wires for the stimulating electrodes exit from the titanium capsule through a ceramic seal in its base. The lead wires then pass to the periphery and emerge from the front of the encapsulated

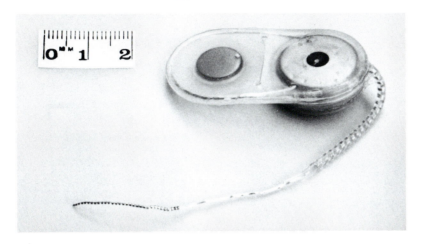

Fig. 47. Photograph of the smaller receiver-stimulator with magnet for implantation in children. ×1.2.

receiver-stimulator, where they are directed downwards at an angle of 45 degrees. After emerging from the encapsulated receiver-stimulator the lead wires spiral to reduce the effect of stresses on the wires. The dimensions of the lead wire assembly are the same as for the adult device. The spiralled section is 35 mm in length. The remaining section, which is 30 mm, terminates in the electrode array which has 22 platinum bands for multiple-channel stimulation, and 10 bands unconnected to wires and used for mechanical support. The 22 active electrodes are around the terminal 17.5 mm of the array, and the 10 supporting bands the remaining 7.5 mm. The platinum receiving coil for power and coded speech signals also exits through the ceramic seal in the base of the capsule, but emerges from the back of the encapsulated receiver-stimulator opposite the lead wires for the stimulating electrodes. The coil extends outwards and backwards from the capsule containing the electronics. It does not lie around the package as is the case with the adult device. Placing the receiving coil behind the electronics capsule allows the thicker section containing the electronics capsule to be placed more anteriorly in the mastoid beneath the pinna if this is surgically desirable. The receiving coil then lies behind the pinna, and because it is flatter no obvious swelling results. The receiv-

Fig. 48. Photograph of the laboratory computer used for the software speech processor developed to help profoundly deaf patients understand running speech.

ing coil also surrounds a rare earth magnet which is sealed in a titanium case. This implanted magnet enables the externally worn transmitter coil, which also contains a magnet, to be easily attached and aligned by the child. The receiver-stimulator capsule, coil and magnet are encapsulated in Silastic MDX-4-4515, as a flat package which has a maximum length of 46 mm, and breadth of 24 mm. The height of the section containing the titanium capsule is 6 mm, and the section with the coil is 3 mm.

In making a smaller device it was also decided not to incorporate a connector. This decision was based on a number of factors. Firstly, failure modes although rare may not be in the receiver-stimulator package, and could occur in the connector or electrode array. Secondly, our laboratory studies on animals have shown that the free-fitting, smooth, banded and tapered electrode array can be explanted and another reinserted, with minimal or no trauma. Thirdly, our clinical studies on patients have shown no deleterious effects on their performances with explantation and reinsertion.

Laboratory-Based Speech Processor

A speech-processing strategy that would help our profoundly-totally deaf patients with a postlingual hearing loss understand running speech was first established [36] using a laboratory-based computer (fig. 48).

Laboratory-Based Speech Processor

With this speech-processing strategy four speech parameters were extracted from the acoustic wave form. These parameters were the fundamental or voicing frequency (F0) and its amplitude (A0), and the second formant frequency (F2) and its amplitude (A2). The F2 information was allocated to selected electrodes; the electrodes selected were on the basis of results obtained by ranking electrode sensations from sharpest to dullest. Each electrode was assigned to a particular portion of the F2 frequency range whereby the electrode with the dullest sensation or lowest pitch was assigned to the lowest frequency proportion, while that with the sharpest sensation or highest pitch was assigned to the highest frequency portion. The electrode was stimulated by a pulse train determined from the F0 parameter. The speech segment was classified as voiced if A0 exceeded a preselected threshold, otherwise it was classified as unvoiced. For unvoiced speech the pulse rate was low and random, while for voiced speech the rate was proportional to the F0 frequency. The current level at which the pulse train was presented to the electrode was determined by A2. The upper limit at which the electrode was stimulated was set at a current level that did not exceed a comfortable loudness for the patient.

This speech processing strategy helped our first patient understand some running speech when using electrical stimulation alone, and all audiological tests showed big improvements when electrical stimulation was combined with lip-reading compared to lip-reading alone [39, 40]. Similar encouraging results were obtained from a second patient operated on in July 1979 [39, 40].

Portable Prototype Speech-Processor

The next task was to design a portable prototype speech processor that would allow the patient to use the speech-processing strategy in an everyday situation. This task was completed towards the end of 1979, and the portable unit which was about the size of a binocular case is shown in figure 49 being used by our first patient to hold a conversation. The

Fig. 49. Photograph of the portable prototype speech processor developed by the University of Melbourne.

speech processor had dimensions of $15 \times 15 \times 6.5$ cm, and a total weight of 1.25 kg. It was constructed from linear and CMOS integrated circuits. It dissipated approximately 2W of power, and was powered by a NiCd rechargeable battery. The patient wore a head-set placed over the vertex to hold the transmission coil in place, and used a hand-held microphone.

It was established that the portable prototype speech processor could help postlingually deaf patients understand running speech [131, 132]. For this reason a public interest grant was made by the Australian Government for its industrial development. The biomedical firm Nucleus Limited whose subsidiary Telectronics had played a major role in cardiac pacemaking, tendered and was awarded the contract.

Wearable Speech Processor for Clinical Trial
The receiver-stimulator manufactured for clinical trial has already been discussed above. The speech processor produced by Nucleus Limited or its subsidiary Cochlear Pty Limited is shown in figure 50.

With the speech professor, speech signals are picked up by a directional microphone worn on the headset above the ear. This converts acoustic sounds or speech into electrical signals for processing. For cochlear implant patients it is important that a directional microphone is used. This is because patients are presently implanted monaurally and it is well-

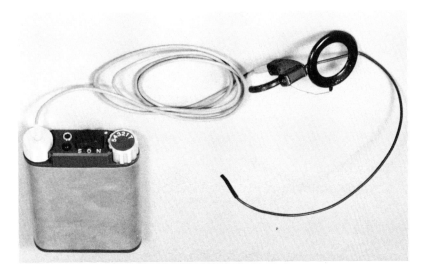

Fig. 50. Photograph of the wearable speech processor and headset developed by Cochlear Pty Limited for clinical trial by the FDA.

documented that monaural processing of auditory information is much less effective in noisy environments. The directional microphone enhances sound signals coming from the front and reduces those from other directions. As a result, if the microphone is pointed towards the speaker, the speech signal in a noisy environment will be enhanced. Another consideration for microphone selection is that it should be small to enable ease of carrying and for cosmetic reasons.

The amplified speech signal is directed along three parallel paths for the extraction of the following three speech parameters: amplitude envelope, fundamental frequency (F0) and formant frequency (first formant F1 and second formant F2). The output of the amplitude envelope detector is converted to electrical current level for the speech processor extracting F0 and F2, but with the newer speech processor extracting F0, F1 and F2 the amplitudes of the first and second formants (A1 and A2) are extracted and converted to current level. The estimated fundamental frequency is converted to electrical pulse rate and formant frequency to electrode position. These electrical parameters are fed into the output and configured for transmission at radio frequencies to the implanted receiver-stimulator. This is illustrated in the block diagram in figure 51.

An automatic gain control amplifier reduces the range of amplitude

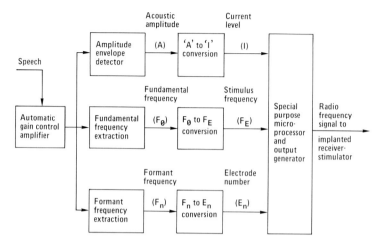

Fig. 51. Block diagram of the speech processor [50].

values (dynamic range) of the speech sounds to be processed both by the device and for the benefit of the patient. The amount of dynamic range reduction and the time delay over which this reduction is achieved for ongoing speech are factors which must be carefully controlled to ensure a minimum amount of distortion of the speech signals. The amplitude envelope detector consists of a full-wave rectifier followed by a low-pass filter.

Electrical current has been found in psychophysical experiments to be related to acoustic amplitude by a power function and the power coefficient can be determined by psychophysical means. It should be noted that the power coefficient may vary from electrode to electrode depending on the size and location of the electrodes, and on the density of residual nerve fibres in the cochlea. Directed by the controlling logic, the acoustic amplitude is sampled by an analogue-to-digital converter. A look-up table prepared according to the functional relationship is stored in a block of digital memory from which a current level can be read once the acoustic amplitude is sampled and specified. Fundamental frequency is measured by passing the output of the automatic gain control segmentally through a rectifier, a low-pass filter, a zero crossing counter and a frequency-to-voltage converter. An analogue voltage proportional to the fundamental frequency is produced at the output of the frequency-to-voltage converter. This voltage can be optionally scaled to represent an electric frequency which falls in the best range of electrical frequency discrimination. The formant frequency is estimated from the output of a formant filter which is designed to cover the frequency range of the formant frequencies in question. The filter output is fed through a zero crossing counter and frequency-to-voltage converter, to produce a voltage proportional to the formant frequency. Directed by logic, this voltage is sampled by an analogue-to-digital converter and the electrode number read from a look-up table in digital memory. The correspondence between formant frequency and electrode number can be determined by psychophysical studies, or derived from the basis of the spacing between the electrodes in the cochlea and known physiological principles.

The signal parameters described above are continuously fed to the controlling logic and output configuration in real time. A CMOS gate array encoder controls the instants at which the parameter values are sampled and configures the radio frequency to be transmitted to the implanted receiver-stimulator. The speech processor shown in figure 50 has been implemented using standard integrated circuits in hybrid form with CMOS technology. Speech information that has been transformed into electrical signals is transmitted to cochlear electrodes to excite residual auditory nerves. In addition electrical power is also provided. Speech data and power are both transmitted by electromagnetic coupling using a carrier frequency of 2.5 MHz. Electromagnetic induction is used as it is the most efficient and reliable method of transmitting the signals. The receiver coil which is outside the protective hermetic enclosure is a single turn of platinum wire, while the external transmitter coil has eight turns. The transmission has been designed so there is adequate power transferred over a distance of 10 mm when the coils are coaxial and some degree of misalignment is possible in shorter distances.

Diagnostic and Programming System

A diagnostic and programming system (DPS) is used to program a patient's wearable speech processor with the threshold, comfortable listening and maximum discomfort levels for individual electrodes. This information is obtained from psychophysical tests using the diagnostic and programming unit, and is written into an erasable programable read only memory (EPROM) chip in the form of a reference map. When speech is processed by the wearable speech processor the appropriate

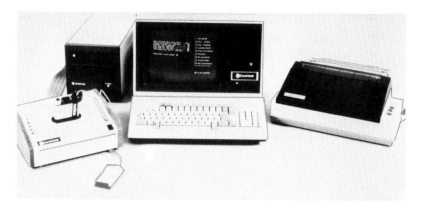

Fig. 52. Photograph of the diagnostic and programming system used to establish and programme thresholds, comfortable listening levels and maximum discomfort levels for each electrode onto the map in a patient's wearable speech processor.

stimulus parameters for each electrode are obtained by referring to this map. The DPS for use in patient evaluation is shown in figure 52.

The DPS consists of a custom designed microprocessor based speech processor interface (SPI) and an off-the-shelf microcomputer referred to as the diagnostic and programming unit (DPU) for control of the SPI. The DPU communicates with the SPI with a high-speed bi-directional parallel communications link, and the communications protocol includes extensive error checking for avoiding delivery of unwanted stimuli to the patient. A block diagram of the DPS is part of the whole system and is shown in figure 53.

When being tested the patient's own wearable speech processor is plugged into the SPI, which then assumes control of the wearable speech processor. The SPI is able to cause the wearable speech processor to encode different stimulus parameters allowing a wide variety of psychophysical experiments to be performed. Once a map is generated, this can be tested with 'live speech' from a microphone and, if found to be satisfactory, can be programmed into the EPROM. The SPI includes an ultraviolet lamp for erasure of the EPROM to allow reprogramming. Analogue inputs are used for reading a patient knob which is used for a variety of experiments, and for allowing the patient to adjust stimulus levels when measuring thresholds and comfortable levels. The SPI includes a stimulator which takes as its input the signal transmitted from

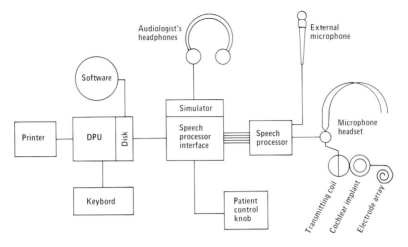

Fig. 53. Block diagram of the diagnostic programming system.

the patient's head coil; a visual display shows in real-time, the stimulated electrode, and an audible signal is generated in a pair of headphones for each stimulus pulse. This stimulator has proved to be extremely useful in checking problems recorded by patients. The DPU is an IBM PC or SANYO microcomputer with a custom interface to the SPI. Several man years of work have produced a suite of programmes which allow great flexibility in conducting psychophysical experiments and configuring the patient's map. Data from tests are stored on flexible disks, which gives a convenient method for archiving and communicating test results. The DPS programmes have five main sections: threshold measurement; loudness scaling; place-pitch ranking; pulse rate discrimination, and map generation. The measuring threshold section is used to measure thresholds and comfortable levels for each electrode. A burst of stimulus pulses, at a rate fixed by the audiologist (usually 125 Hz), is delivered to the selected electrode. The stimulus level for each presentation may be set from the keyboard by the audiologist, or may be determined by the patient himself by using the patient knob.

We have defined threshold to be that level whereby the patient can always just hear the signal. Comfortable level is defined as the highest level which the patient can listen to for a long time without it being uncomfortable. These measurements are important as they define the minimum and maximum stimulus levels which will be programmed into the patient's map in the wearable speech processor, and the range over which the loudness transfer function is to be calculated. Restricting stimuli within this range guarantees that the patient will not be delivered uncomfortable stimuli. Unlike a single-channel cochlear implant, the use of a multiple-channel hearing prosthesis allows exploration of the tonotopic arrangement in the cochlea. This is referred to as place-pitch to distinguish it from rate-pitch. Stimulation of more apical electrodes yields percepts recorded as dull or low place-pitch, while stimulation of basal electrodes is reported as sharp or high place-pitch. As might be expected an essentially tonotopic arrangement is recorded and patients are generally able to reliable distinguish between adjacent electrodes.

Selection of Patients

Patients who are now selected for implantation by the Cochlear Implant and Rehabilitation Clinic at the Royal Victorian Eye and Ear Hospital must meet certain criteria. They should have a profound-total hearing loss, be postlingually deaf, have no psychiatric contra-indications and an intelligence quotient in the normal range, have no oto-logical or X-ray findings to contra-indicate implantation, have positive results for electrical stimulation of the promontory, and be medically fit for the surgery. These criteria are summarized in table 10.

Profound-Total Hearing Loss

Patients who are selected for cochlear implantation in our clinic should have a profound-total hearing loss and receive no significant benefit from a powerful hearing aid. A profound-total hearing loss has been defined as an average pure tone threshold greater than 90 dB HL in the better ear with the frequencies 500, 1,000, and 2,000 Hz. The patients are tested preoperatively using the most effective aid. Aided free-field thresholds are measured for pure tones at 0.5, 1.0, 2.0, and 4.0 kHz. The most important preoperative finding, however, is not the pure tone threshold, but whether the patient can achieve useful communication with a hearing aid or tactile device, and whether it is better than expected from a cochlear prosthesis. This is best confirmed using open sets of pho-netically balanced words and the Central Institute for the Deaf (CID) Everyday Sentences [61] in the aid-alone condition. In the present state of knowledge, prospective patients should have zero open-set word scores when tested under the above conditions. This principle is applied even though results with the clinical trial of our formant based multiple-channel prosthesis have shown a high percentage of patients obtaining

Table 10. Selection criteria

1	Profound-total hearing loss
2	Postlingual deafness
3	Psychiatric state and IQ
4	Otology
5	X-Rays of skull including cochlea
6	Electrical stimulation of promontory
7	Medical condition

postoperative open-set CID Everyday Sentence scores above zero when using the implant alone. For example their scores have ranged up to 86%. A battery of audiological tests is also used to compare the performance of the two ears. This battery of tests, which is shown in table 11, is also used to assess the performance of the patient postoperatively.

Postlingual Deafness

The definition of when a patient is prelingually or postlingually deaf has not been generally agreed upon. It is important, however, that whatever criterion a centre uses it should be clearly stated, so that the results between patients and between centres can be compared. Studies have shown that although the acquisition of language can vary between individuals, it may not be complete until 6 years of age. In our studies on postlingually deaf adults we have only included patients if they lost hearing after the age of 4 years.

Psychiatric Status and Intelligence Quotient

A severe psychosis or psychoneurosis not directly related to the deafness is a contra-indication to surgery. It is important that the patient has the right mental state to cooperate with the rehabilitation team otherwise results are not likely to be optimal. The IQ has not been used to select patients if it is within the 95% level. In fact the IQ level has not been found to correlate with patient results.

Table 11. Preoperative and postoperative tests

1   Thresholds
    Signal detection level
    Maximum comfort level
    Loudness discomfort level
2   Minimal auditory capabilities battery (preoperative auditory or postoperative electrical stimulation alone)
    One versus two syllables
    Spondee same/different
    Noise/voice
    Accented word
    Male/female speaker
    Question/statement
    Four-choice spondee
    Vowels
    Initial consonants
    Final consonants
    Everyday sounds (closed-set)
    Everyday sounds (open-set)
    Spondee recognition
    CID everyday sentences
    NU monosyllable words
3   Assessment of the value of aid preoperatively or cochlear implant postoperatively
    for lip-reading
    CID everyday sentences
        Aid/cochlear implant plus lip-reading
        Lip-reading alone
    CNS monosyllabic words
        Aid/cochlear implant plus lip-reading
        Lip-reading alone
4   Additional postoperative tests
    Vowels and consonants
        Electrical stimulation plus lip-reading
        Electrical stimulation alone
        Lip-reading alone
    CID everyday sentences
        Electrical stimulation alone
    Speech tracking
        Electrical stimulation plus lip-reading
        Electrical stimulation alone (when patient's performance
        permitted)
        Lip-reading alone

Otology

The otological history and examination is very important prior to cochlear implantation. It is necessary to exclude active infection in the external or middle ear and to identify perforations of the tympanic membrane or previous middle ear or mastoid surgery. A cerebellopontine angle tumour should also be excluded. If active infection is present in the external auditory canal, middle ear or a mastoid cavity, this must be treated before proceeding with an intracochlear or extracochlear implantation. This is especially important for the intracochlear insertion of an electrode array otherwise a labyrinthitis is very likely to occur. Furthermore, extracochlear implantation where bone is drilled should also be avoided as infection is equally likely to occur and may also enter the inner ear.

X-Rays

X-Rays are one of the most important tests requires before proceeding with the operation. A plain X-ray will show the degree of aeration of a mastoid cavity and exclude any obvious disease. It is also useful in the identification of large emissary veins that could cause severe bleeding at the time of surgery. A polytome radiograph or CT scan is, however, most useful in selecting patients for multiple-channel cochlear implantations. Best views are now obtained with a CT scan. An example of a normal cochlea with patent basal turn and a round window niche is shown in figure 54. The CT scan is especially useful in showing whether the round window niche is present and the extent of bony obliteration of the first part of the basal turn. In about one third to one half of our patients there has been disease of the basal turn, and the CT scan is very helpful to the surgeon in knowing how much drilling is required before exposing the basal turn and in giving him the assurance that, following the drilling, an electrode will pass adequately into the cochlea. In only one case out of 55 operations has it not been possible to pass a multiple electrode array. In this particular patient, meningitis had occurred. The polytome X-ray showed that the basal turn was probably patent, but at surgery it was found to be filled with fibrous tissue. This fibrous tissue was, however, more easily recognizable on a CT scan. This is shown in figure 55. In this particular patient a CT scan showed the opposite ear to be patent and an

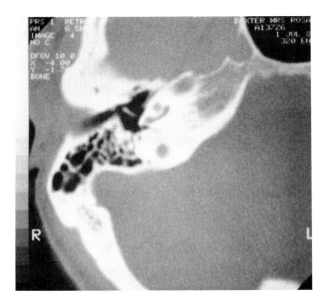

Fig. 54. CT scan of the cochlea of a profoundly-totally deaf patient showing a normal cochlea and the presence of a round window niche.

intracochlear multiple-electrode array was subsequently introduced with good results in the opposite ear. Finally it is important to emphasize that, in a small proportion of cases, patients presenting for implantation have grossly diseased cochleas; for example, cochlear otosclerosis. In these cases it is important to recognize that occasionally two pathologies can occur. This is illustrated in figure 56. This patient presented with advanced cochlear otosclerosis but also had, at the time, radio translucency due to a Schwannoma of the vestibular nerve. This subsequently increased in size, caused deterioration in results, and finally needed to be removed.

Electrical Stimulation of the Promontory

All patients who are considered suitable for a cochlear implant have electrical stimulation of the promontory. Our first 10 patients had stimulation with biphasic pulses with pulse widths of 200 µs, and the

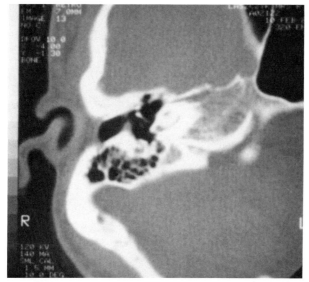

Fig. 55. CT scan of the cochlea of a patient who lost hearing following meningitis. *a* The basal turn on one side was filled with fibrous tissue which prevented intracochlear implantation. *b* Implantation was carried out on the other side which was patent.

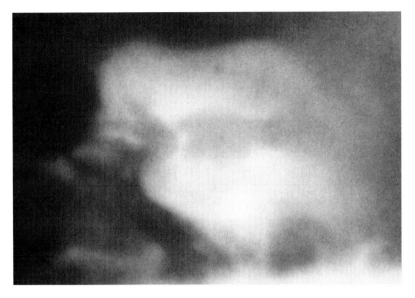

Fig. 56. Polytome of the cochlea of a patient who had advanced cochlear otosclerosis. Radio translucency can also be seen due to a Schwannoma of the vestibular nerve.

remainder had a continuous square-wave stimulus. The former pulse wave form resulted in more patients experiencing pain.

When carrying out the tests, patients are asked to report on the sensations, and whether they are heard, felt or painful. They are then encouraged to describe the sounds and liken them to any sound they could previously remember. Thresholds from 50 to 1,600 pulses/s are recorded, and rough estimates of difference limens for rate established when possible.

Medical Condition

The general medical condition of the patient should be sufficient to allow the operation to be carried out safely. As the patient has a lot to gain from the procedure in the same way as a patient having a cataract operation, a similar selection criterion should be used.

Factors Predicting Patient Benefits

To determine the factors that are important in selecting patients most likely to benefit from a multiple-channel cochlear prosthesis, we have carried out a statistical analysis to compare patient results with age, duration of profound hearing loss and numbers of usable stimulus channels. A multiple regression analysis showed no correlation between age and performance indices, a highly significant negative correlation between the length of profound deafness and performance, and a highly significant positive correlation between the number of usable channels and performance [67]. With regard to the duration of deafness, the performance was worst if the hearing was lost more than 13 years before surgery. Furthermore it was of interest that three patients who each had a profound hearing loss of more than 30 years had vowel recognition scores which were above the average. They were receiving an amount of formant information similar to the other patients. On the other hand, their scores for open-set sentence tests and improvements in speech tracking were significantly lower than the average, indicating that when they were presented with the same information as the others, they were not able to use it as effectively in the recognition of running speech. This was possibly due to a loss in central auditory processing, resulting from the long period of sensory deprivation. These patients also took longer to reach a plateau in their performance levels than the others.

The Surgery

The scala tympani is a convenient site to place an electrode array close to auditory nerve fibres. In this scala the electrodes lie below the basilar membrane and the dendrites running through the spiral lamina to the ganglion cells. The Nucleus electrode array is 25 mm long. It has 22 active banded electrodes spaced every 0.75 mm from the tip, and 10 inactive bands for mechanical support. It is not, however, necessary to place the entire array into the scala tympani in order to have a system which will function effectively for the patient. It must be remembered that the 10 most proximal bands, extending over a distance of approximately 7.5 mm, cannot be activated, and function only to provide better mechanical characteristics to facilitate the array being placed into the scala. Typically, the total length that can be inserted into the scala tympani varies from 15 to 25 mm depending on the anatomy and pathology. In some cases, given that resistance is felt, fewer electrodes are inserted. In such a case, the electrodes which are not located in the scala tympani are not used for stimulation, and are inactivated during the initial programming of the device. The important factor is that the electrodes pass smoothly into the scala with minimal trauma.

The cochlear implant consists of an electronics package which provides the means by which externally transmitted signals are decoded and transmitted to the electrode array, an antenna which receives the external signal by radio waves transmitted through the skin and subcutaneous tissues from a corresponding antenna on the skin surface and the electrode array containing the 22 active bands and 10 mechanical support bands. The receiver-stimulator and antenna are mushroom-shaped.

The receiver-stimulator is placed in a bed cut into the mastoid (and possibly the parietal or occipital bones). The electrode lead passes from the receiver-stimulator to the round window via the mastoid cavity and a posterior tympanotomy. The electrode is inserted as deeply as possible so that a maximum number of nerve fibres around the basal turn can be stimulated. Excessive force must not be used in the insertion to avoid perforation of the basilar membrane or fracture of the spiral lamina both of which may result in a localized loss of spiral ganglion cells. The bundle of lead wires which lie between the package and electrode array is secured to bone using ribbons of Dacron® mesh. The round window is sealed with a plug of connective tissue. The wound is closed in layers so that the receiver-stimulator is completely covered with flaps of deep fascia, subcutaneous tissue and skin. When the wound is well healed, usually at 2–3 weeks postoperatively, the implant can be activated.

The aims of the surgery are, firstly, to aseptically and atraumatically implant an electrode array adjacent to the auditory neurons so that discrete groups can be stimulated electrically, and secondly to provide a secure placement of the associated package beneath intact skin. This package must be situated so that it can be stimulated readily by radio waves from an overlying transmitter antenna. The stages in the surgery are summarized in table 12.

Preparation of Patient

The operation is performed under general anaesthesia, with the patient receiving broad spectrum antibiotic coverage, commencing 2 h prior to surgery. Access to the operating room is limited to necessary personnel in order to minimize the chance of contamination. A strict aseptic procedure is used and it is desirable to operate on a horizontal laminar flow of filtered air [38]. The sterile package containing the cochlear prosthesis is not opened until the prosthesis is actually to be inserted.

In the anaesthetic room hair is shaved in a generous area, in order to allow sterile draping of the ear, mastoid, parietal and occipital areas. The patient is placed supine on the operating table and the head turned away from the ear to be operated upon. The shaved area is prepared with Betadine solution and a sterile plastic drape is applied. The remainder of the body is covered with standard operating room drapes.

The Incision

A dummy receiver-stimulator is positioned so that its upper edge is level with the superior attachment of the auricle, and the anterior edge is approximately 1 cm behind the back of the auricle.

Table 12. Surgery

---

1 Preparation of patient
2 Incision
  A The inverted U-incision
  B The C-incision
3 Mastoidectomy
4 Posterior tympanotomy
5 Receiver-stimulator package bed
6 Preparation of the groove joining the package bed to the mastoid cavity
7 Preparation for the insertion of the electrode array
  A The hook region
  B Preparation of the round window opening
  C Entering the scala tympani
8 Placement of Dacron tape
9 Electrode insertion
10 Stabilization of the implant
11 Facial flap closure
12 Skin closure
13 Surgery for implantation of smaller mini receiver-stimulator for children

---

Either inverted U- or C-shaped incisions can be used. With a marking pen a flap is outlined, the margins of which are at least 2 cm from the edge of the dummy in all directions. Whichever flap is used, infiltrate the flap site with a vasoconstricting agent. Inject along the line of the incision, beneath the flap and into the posterior skin of the external auditory canal.

The Inverted U-Incision

The dimensions of the flap should be large enough to permit flexibility in positioning the receiver-stimulator package and the base should be wide enough to maximize the vascular supply to the skin at its periphery (fig. 57). It is desirable that no part of the skin incision lie over the package as this could lead to dehiscence or interfere with the proper application of the external coil, which must lie directly over the internal antenna. After the inverted U-shaped flap of skin and subcutaneous tissue is created, a second flap composed of deep fascia and periosteum is fashioned. The second flap is based anteriorly (fig. 58).

The creation of two separate flaps is desirable as it could reduce the likelihood of a superficial wound infection extending inward around the

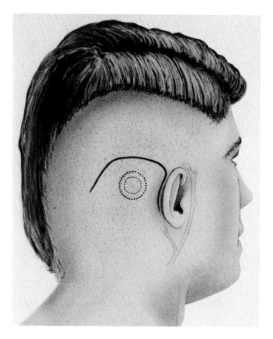

Fig. 57. Cochlear Corporation – Surgeon's Manual. The inverted U-shaped incision.

implanted package. It is also important that the two incisions do not over-
lie each other or the package, as a fistula could otherwise result or even
partial extrusion of the package occur.

The flap of skin and subcutaneous tissue is elevated by dissecting in
the plane between the subcutaneous and deep fascia. Haemostasis is
achieved, the flap wrapped in a warm moist cloth, and the self-retaining
retractors inserted.

The incision for the anteriorly based flap of deep fascia and periosteum is then made. It begins at the posterior root of the zygoma and
normally continues along the supramastoid crest, but must be not less
than 1 cm internal to the skin incision. The posterior margin of the incision also runs parallel to and 1 cm internal to the skin incision. The
inferior margin of the incision runs along the superior nuchal line to the
mastoid tip. If the incision for the inferior margin is made too low, the
occipital artery may be encountered and require ligation.

The deep fascial and periosteal flaps are then elevated to expose the
mastoid bone, the mastoid process, the suprameatal spine, the supra-

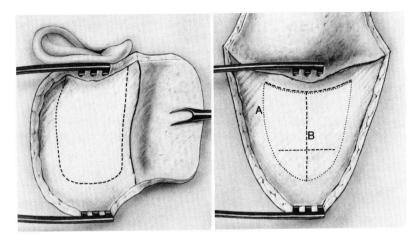

Fig. 58. Cochlear Corporation – Surgeon's Manual. Left: Inverted U-shaped flap of skin and subcutaneous tissue elevated to expose the deep fascia. The outline of the anteriorly based flap of deep fascia and periosteum is demonstrated. Right: C-shaped flap of skin and deep fascia elevated to expose the deep fascia. Two additional methods of incising the deep fascia are shown.

meatal triangle and the posterior bony wall of the external auditory canal. Elevation of the periosteum may cause bleeding from a mastoid emissary vein which needs to be controlled.

The C-Incision

The flap is designed with diverging superior and inferior borders in order to assure an adequate blood supply via the superficial temporal artery above and the postauricular artery below. There should be at least 2 cm between the receiver-stimulator and the incision in all directions (fig. 58, 59).

The Mastoidectomy

A mastoidectomy is carried out exposing the mastoid antrum. Overhang is allowed to remain superiorly and inferiorly, through which holes can be drilled for later insertion of the ties used to attach the thick lead

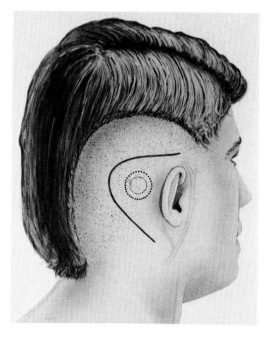

Fig. 59. Cochlear Corporation – Surgeon's Manual. C- shaped incision.

wire from the receiver-stimulator package firmly to bone. The excavation must be adequate for a safe posterior tympanotomy and good exposure of the round window niche. A groove is finally drilled between the receiver-stimulator bed and the back of the mastoid cavity. This is undercut to protect the lead wire bundle.

Begin by removing the mastoid cortical bone over the suprameatal triangle. When the mastoid antrum is reached, the exposure is widened so that both the lateral semicircular canal and the short process of the incus can be clearly seen. Extend the cell removal inferiorly to the digastric ridge. Preserve the bone over the sigmoid sinus and if possible do not bevel off the posterior edge of the cavity.

It is desirable to retain as much bone over the sigmoid sinus as possible, so that the package bed can be created with an anterior wall that will prevent the receiver-stimulator unit slipping forwards. If, however, the sigmoid sinus is more anterior and superficial than normal, some of the overlying bone will need to be removed in order not to compromise the approach to the posterior tympanotomy.

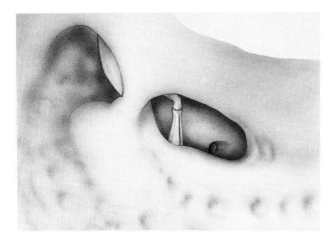

Fig. 60. Cochlear Corporation – Surgeon's Manual. Posterior tympanotomy showing the superstructure of the stapes, part of the long process of the incus and the round window niche.

In widening the mastoid cavity, the posterior external canal wall will also need to be thinned in order to obtain a good access to the round window through the posterior tympanotomy. If a hole is made, then it should be covered with cartilage at the end of the procedure to prevent the possibility of a fistula.

The Posterior Tympanotomy

In carrying out the posterior tympanotomy ensure that the key landmarks, the lateral semicircular canal, the short process of the incus and the anterior end of the digastric ridge are easily seen (fig. 60).

Demonstrate the facial nerve and retain a thin layer of bone on its anterior and lateral surfaces as the posterior tympanotomy is prepared. Preserve the chorda tympani, if possible. Retain the bridge for the incus, and proceed with the tympanotomy to the point where both the stapes suprastructure and round window are visible.

In demonstrating the facial nerve, drill along the line between the short process of the incus and the anterior end of the digastric ridge. Aim to demonstrate the facial nerve on its posterior and lateral aspect before proceeding with the posterior tympanotomy. Always move the drill along the line of the nerve and watch for any angulation laterally and posteriorly. The nerve should be seen and its course confirmed from a point just inferior to the fossa incudis to a point 1 cm towards the mastoid tip or the junction of the chorda tympani with the facial nerve. It is essential to avoid damage to the facial nerve, yet provide a generous posterior tympanotomy without entering the ear canal or injuring the chorda tympani, if possible.

In completing the posterior tympanotomy drill the bone away, lateral to the skeletonized facial nerve, keeping a thin plate of bone over the nerve laterally and anteriorly, to protect it from the shank of the burrs. Deepen a channel anteriorly between the facial nerve and the chorda tympani until the facial recess of the middle ear cavity is entered. Make the site of entry at the upper end, beneath the floor of the fossa incudis. An air cell sometimes situated at this point may facilitate the entry. Then remove the bone in the depths of the tympanotomy posteriorly until the round window niche is visible. The posterior tympanotomy should be at least 2 mm wide for an adequate exposure of the round window. In some cases, the fibrous annulus of the tympanic membrane will have been exposed to gain an adequate view (fig. 60).

The Receiver-Stimulator Package Bed

The receiver-stimulator package bed should be made so that the package will be stable, and not slide forward or rock. The ring-shaped template is sited over the mastoid region (fig. 61). The front edge must be 1 cm behind the postaural sulcus. The centre of the package should be not too high to encroach on the thin bone over the squamous temporal bone and not too low to rock on the curved part of the skull behind the mastoid process. Use a 2-mm cutting burr around the inside of the ring to delineate the site of the bony excavation.

Using the cutting burrs a circular bed is then drilled to fit the receiver-stimulator (fig. 62), and its sides and bottom are smoothed using the special diamond milling burrs (fig. 63). The siting of this bed is made by releasing the retractors and positioning the dummy package so that the

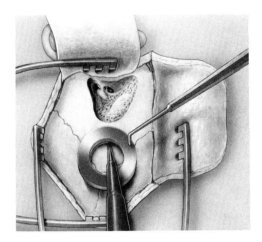

Fig. 61. Cochlear Corporation – Surgeon's Manual. Completed mastoidectomy and posterior tympanotomy. The template is in place to delineate the site for drilling the bed to take the stem of the receiver-stimulator.

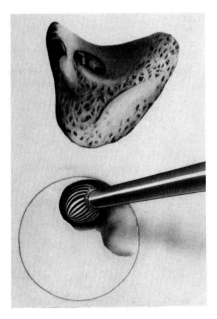

Fig. 62. Cochlear Corporation – Surgeon's Manual. Bed for the receiver-stimulator being drilled with a cutting burr.

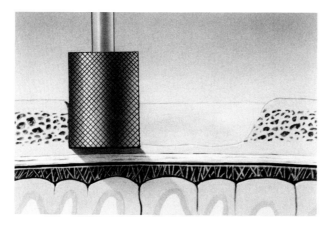

Fig. 63. Cochlear Corporation – Surgeon's Manual. Milling burr is used to cut the edges of the bed perpendicularly, and to smooth the floor.

anterior edge lies at least 1 cm behind the postaural sulcus. During the drilling, dura may be exposed. The mastoid emissary vein can be encountered and bleeding can usually be controlled with bone wax. Two holes may be drilled on either side of the bed in order to accommodate the receiver-stimulator package stay sutures, later to be used to secure the receiver-stimulator in its bed. After removing the bone and completing the package bed, the disc-shaped template is used to ensure that the bed is just larger than the package, so that a tight fit will be obtained.

Preparation of the Groove Joining the Package Bed to the Mastoid Cavity

It is necessary to prepare a groove flush with the floor of the bed to join the bed to the mastoid cavity (fig. 64). The groove should have overhanging walls to provide protection and holes for the Dacron mesh ties (fig. 64). The groove receives the first 2–3 cm of electrode lead wire bundle. The groove may be placed superiorly so that the lead wire runs towards the mastoid antrum or inferiorly so the wire passes around the mastoid tip. Drill away any sharp corners on the groove and place two pairs of holes in the overhang, as shown in figure 64.

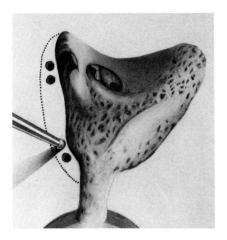

Fig. 64. Cochlear Corporation – Surgeon's Manual. A gutter is drilled between the receiver-stimulator bed and the mastoid cavity. Pairs of holes are drilled in the bone overhanging the groove for the placement of the Dacron ties used to secure the thick electrode lead wire bundle.

Preparations for the Insertion of the Electrode Array

The Hook Region

An understanding of the anatomy of the hook region of the basal turn of the cochlea is critical for the atraumatic insertion of the electrode array. The hook results from a widening of the cochlear spiral as it approaches the round window (fig. 65), and invagination from a crest of bone, the crista fenestrae, which lies inside the antero-inferior margin of the round window (fig. 65). The crista fenestrae obstructs the view along the basal turn. If it is not drilled away with a 0.6-mm diamond burr, the electrode can be directed towards the modiolus and this may either damage the basilar membrane or result in the electrode coiling on itself.

Preparation of the Round Window Opening

Visualize the stapes to confirm the site of the round window. It can be confused with an air cell in the hypotympanum (fig. 66). Drill away the bony overhang, if this has not already been done. Care must be taken if it is necessary to drill posterosuperiorly to avoid any risk of damaging the underlying spiral lamina. The drilling should take place primarily anteroinferiorly. Please note that a hypotympanic air cell can lie just beneath the

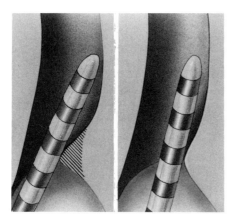

Fig. 65. Cochlear Corporation – Surgeon's Manual. Diagram of the scala tympani of the basal turn of the cochlea showing the importance of removing the crista fenestrae to allow the electrode array to be inserted in the right direction.

round window and be mistaken for it, especially when the round window is obliterated (fig. 66).

Exposing the Scala tympani

If the round window membrane is normal, it is possible to enter the scala tympani by reflecting it from its antero-inferior bony margin (fig. 67). It is important to obtain a good view along the basilar turn as this helps ensure that the electrode array will be inserted without damage to the basilar membrane. Expand the opening to 1 mm in diameter. It will often be necessary to reduce the crista fenestrae whether the opening is through the window or via a fenestration antero-inferior to the round window. Remove bone dust using diamond paste burrs (0.6 and 1.0 mm) at slow speeds with continuous irrigation.

If the round window membrane and niche are partly or completely obliterated, fenestrate the promontory overlying the scala tympani of the basal turn by drilling about 1 mm antero-inferiorly to the site of the round window niche (fig. 68). The bone is blue-lined and the endosteal layer picked out with hooks. Keep in mind that the diameter of the largest ring is 0.6 mm and the opening needs to be large enough to accept the electrode.

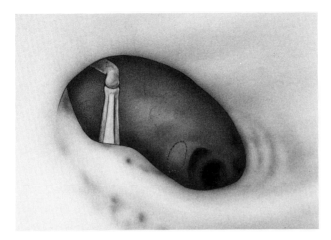

Fig. 66. Cochlear Corporation – Surgeon's Manual. The round window niche can be obliterated and the hypotympanum air cells can look similar to the niche and be mistaken for the it; this is illustrated here.

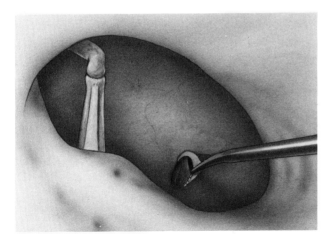

Fig. 67. Cochlear Corporation – Surgeon's Manual. The bone overhanging the round window niche antero-inferiorly may need to be drilled away to expose the membrane. The membrane is incised around its antero-inferior margin. This is illustrated.

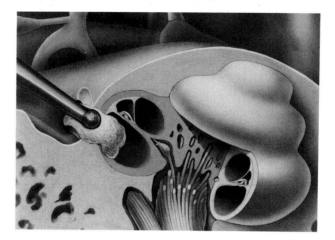

Fig. 68. Cochlear Corporation – Surgeon's Manual. If the round window is obliterated a separate opening can be drilled into the basal turn by fenestrating the bone anteroinferiorly.

In drilling through the bone toward the basal turn keep to the inferior aspect of the promontory and away from the basilar membrane. It is reasonable to continue drilling forward for 5 mm in the case of an obliterated basal turn. Fenestrating the basilar turn antero-inferiorly to the round window may also be used electively with a normal round window niche (fig. 69).

Placement of Dacron Tapes

Thread the tapes through the holes in the bone overhanging the groove for the lead wire. This is facilitated if a small corchet hook is used to pull the tape through the holes. Seat the tapes so that the posterior tympanotomy can be seen through loops (fig. 70).

Electrode Insertion

Take the implant from the container and remove the sleeve from the electrode. Select a pair of claws to be used during the insertion. The

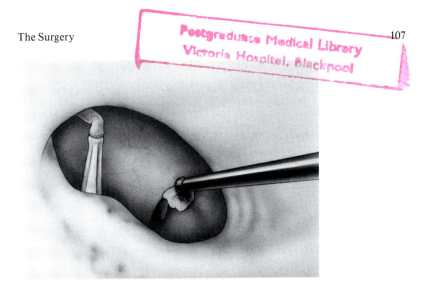

Fig. 69. Cochlear Corporation – Surgeon's Manual. An alternative approach to the round window insertion is via a separate opening drilled anterior-inferiorly to the niche. The bone is blue-lined and the fragments removed with a hook.

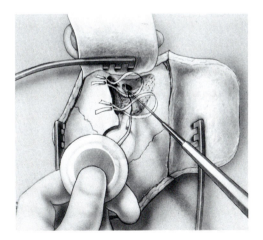

Fig. 70. Cochlear Corporation – Surgeon's Manual. The Dacron ties are in place. The electrode is gently inserted along the scala tympani and this is assisted with special surgical claws.

receiver-stimulator is picked up and held between the thumb and fingers. Try to site the end of the electrode in the round window opening or the fenestration of the promontory by holding the package and touching the lead with the shank of one claw.

Insert the electrode until it passes no further. Usually the electrode moves freely to at least 9–10 mm without assistance from a claw. Guide the electrode further with stroking movements of the claw. The last 10 bands are for strengthening and marking the depth, and do not provide functional stimulation to the cochlear nerve. If the cochlea will not admit those bands, there is little to be gained by a continued attempt to insert the array further into the scala tympani. Never push hard along an electrode if resistance is felt. However, if the electrode passes easily into the scala for the entire 25 mm, it is likely that the patient may be provided with better vowel discrimination as more electrodes will lie close to the fibres which are most sensitive to the speech frequencies. Nonetheless, it is important that insertion should cease when resistance is felt. Two manoeuvres may be used to assist the electrode insertion. However, continued resistance to its forward progress or visible buckling of the array just outside the entrance of the scala tympani strongly indicate that further insertion should not be attempted.

The first manoeuvre is to gently rotate the lead 180 degrees along its length (fig. 28). The rotation should be in the direction of the turn of the cochlea; clockwise in the left ear and counter-clockwise in the right ear. A second manoeuvre is to retract the electrode slightly (1–2 mm) and continue to insert the electrode while gently rotating in the direction of the turn of the cochlea. If the crista fenestrae has been drilled away, or if the entry has been through an antero-inferior opening, there should be no benefit in withdrawing the electrode to reinsert it.

Avoid kinking the electrode lead, engaging the claw on the electrode or burying the claw into the electrode, as damage may occur.

Stabilization of the Implant

When the electrode has been inserted, steady the electrode array in the region of the posterior tympanotomy with a claw. Check that the Dacron mesh tapes do not pull on the lead as the loops are tightened. This can be facilitated by adjusting them. Position the lead as the loops are tightened so that there is no traction of the intra-cochlear aspect of the

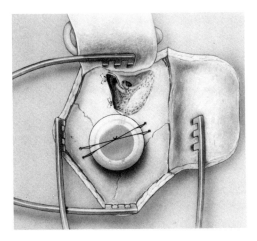

Fig. 71. Cochlear Corporation – Surgeon's Manual. The receiver-stimulator is placed in its bed after the Dacron ties have secured the lead wire in its groove. A tie can also be placed to hold the receiver-stimulator in place.

array. The length of the lead wire is 65 mm. This provides adequate length for a correctly sited implant to lie without redundant loops of lead wire. If the lead wire is slack, it is preferable to have the slack in the part between the round window and the first tie. The second tie is regarded as an important element in tethering the package. Finally place a square of fibrous tissue around the electrode at the opening of the scala tympani. The package can be sutured in place with non-absorbable material (fig. 71). Some surgeons prefer to cross-suture.

Caution. When the implant is in place, it may be damaged by electromagnetic irradiation from diathermy. The diathermy must be used with care near the implant, and only in the coagulation mode and in the bipolar form. The cutting mode must be turned off. Keep the coagulation at the minimum setting which is effective and do not use any closer to the implant than is needed. It is better to ensure that all vessels adjacent to the bed have been controlled before the implant is placed into position. If a vessel is bleeding within 4 cm of the implant, the package should be moved away before the diathermy is used. Take care not to dislodge the electrode when doing this.

Any bleeding vessel which cannot be controlled with the bipolar diathermy should be ligated. Note the strong electromagnetic fields can be generated by the activated diathermy even when the active point does not contact the patient. Accidental triggering of the diathermy, especially in the cutting mode, must be guarded against. Clear the field of all unipolar electro-cautery equipment.

Fascial Flap Closure

Make sure that the package is stable, then bring the fascial flaps up to cover the outer surface of the package or suture around the package. When the flaps are in position, and after good haemostasis is obtained, the junctions are firmly closed with absorbable sutures. This will apply firm pressure to hold the implant in position, and will be better tolerated in the tissues than silk, with less chance of infection. When this closure is complete, irrigate with a dilute antibiotic solution beneath the flaps.

Skin Closure

Check for haemostasis. It is important to control all bleeding as a drain tube should not be used as it is a track down which bacteria can pass to the implant and a haematoma must be avoided because of the risk of infection. Take great care with the diathermy as the risk of damage to the package is not reduced by the fascial cover. In fact, the danger is increased because it is no longer visible.

Take care with the wound closure. The fascial incisions should not lie under the skin incision, so a breakdown of the skin wound will not lead directly to the implant. There should be no gaps and no tension in the sutures, especially in the skin. Interrupted sutures are preferred as a stitch abscess can be controlled readily by a single suture removal.

A closed system drain could also be used and should be placed just inside the periphery of the flap. The incision is closed in layers using dexon for the buried layers and nylon for the scalp. The wound is then covered with a layer of ointment impregnated gauze and a large gentle pressure dressing is applied. The patient is nursed with the upper body and head elevated approximately 30° and kept on antibiotics until the drain is removed 48–72 h postoperatively.

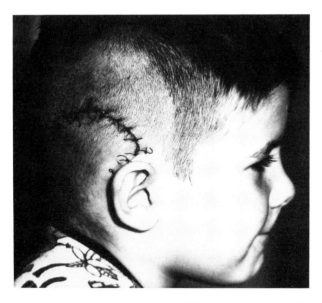

Fig. 72. Photograph of a 5-year-old child 2 days after implanting the small receiver-stimulator unit.

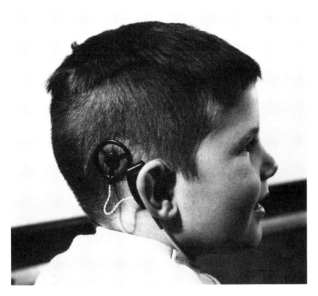

Fig. 73. A child wearing the headset with magnet that is used in conjunction with the smaller 'mini' receiver-stimulator that also has a magnet to help in the alignment and attachment of the transmitting and receiving coils.

The Surgery for Implantation of the Smaller
'Mini' Receiver-Stimulator for Children

The surgery for the implantation of the 'mini' receiver-stimulator is similar to that for the adult device [48], except for the preparation of the package bed. The anterior section containing the electronics capsule can be placed further forwards into the mastoid bowl and beneath the pinna, as it does not have the receiving coil around it. The package bed can, however, be sited in the same place as for the adult. Figure 72 shows a 5-year-old child 2 days after implantation.

The head-set for use with children has been modified to incorporate an external magnet so it can be more easily attached and aligned by the child. A child wearing the head-set is shown in figure 73.

The speech processor for use with children is the same as the one currently available for adults. It can present voicing as rate of stimulation, and the second as well as the first formant as place of stimulation, which is now the preferred method of helping postlingually deaf patients understand running speech [67].

Psychophysics for Postlingually Deaf Adults

Psychophysics has played a very important part in our research to develop a speech processor that would help cochlear implant patients understand running speech. The percepts for different electrical stimulus parameters were analysed to determine the information the patient received. If one knows the percepts obtained with electrical stimulation, a speech-processing strategy can then be developed to code speech in an appropriate way. Furthermore, in training patients in speech perception tasks it is helpful to know if the basic information is being received.

Following the implantation of the prototype multiple-electrode receiver-stimulator in our first patient in 1978, this patient participated in a series of psychophysical tests to determine the perceptual limitations of electrical stimulation. This research led to the development in 1978 of a formant-based (feature extraction) speech-processing strategy that enabled the patient to understand running speech when using electrical stimulation alone, and especially when combined with lip-reading. Psychophysical studies were also undertaken on a second patient who had the prototype receiver-stimulator implanted in 1979.

Further psychophysical research has continued on a larger group of patients implanted with the Nucleus multiple-electrode implant from 1982 onwards. This research has been undertaken to assess the variations in psychophysical performance between patients. It has also been conducted to determine the percepts obtained with more complex electrical stimuli so that improved speech processing strategies could be created. The psychophysical studies and their application to speech-processing strategies are summarized in table 13.

Loudness

Following the implantation of the prototype device in 1978 we, firstly, had to determine the best way to code loudness changes in speech. For this

Table 13. Psychophysics

1   Loudness
2   Pitch
   A   Rate of stimulation
   B   Place of stimulation
   C   Combined rate and place of stimulation
   D   Time varying rate and place of stimulation
3   Speech processing strategy converting voicing to rate of stimulation and second formant to place of stimulation
4   Two component pitch sensation
   A   Two sites of stimulation
5   Speech processing strategy converting voicing to rate of stimulation and first and
   second formants to sites of stimulation

reason we studied loudness as a function of current level and repetition rate, and the results are shown in figure 74. From this it can be seen that the loudness increase with current level was very rapid for electrical stimulation compared to that of the loudness increase with sound pressure level. On the other hand, the loudness increase for pulse rate was much the same as the loudness increase for sound pressure. It is, however, not satisfactory to use rate of stimulation to convey loudness as it was perceived much more strongly as pitch.

As demonstrated above, the range of intensities over which loudness increased to a maximum was very narrow for electrical stimulation. It is fortunate, however, that there were 12–60 discriminable steps within the dynamic range making it possible to use intensity information in cochlear implant speech processing.

Pitch

For speech processing it is not only important to code loudness but even more important to code changes in pitch. Pitch may be coded by rate or timing information as well as place or site of stimulation.

Rate of Stimulation
The psychophysical results for pitch perception as a function of repetition rate are shown in figures 75–78 [135, 137].

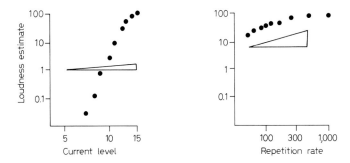

Fig. 74. Loudness estimates for current level and repetition rate ●●●, loudness estimates for sound ———.

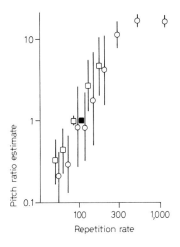

Fig. 75. Pitch ratio estimates versus repetition rate for electrode 1. Multiple-pulses per period stimuli are shown as squares, and single pulses per period as circles. The circles have been displaced to the right by half or one symbol width for clarity. The error bars extend plus and minus one standard deviation of the distribution of log pitch ratio estimates. The solid symbol was the reference stimulus which was 100 pulses/s [135, 136].

Figure 75 shows pitch ratio estimates versus repetition rates. The reference stimulus was 100 pulses/s. Please notice that pitch ratio estimates saturate at 300 pulses/s, indicating that the patient could not discriminate pitches at higher stimulus rates. These results were similar for other patients.

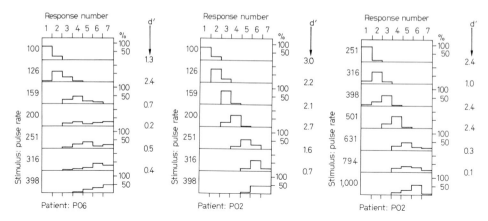

Fig. 76. Confusion matrices in the form of response histograms for the absolute identification of 7 electric pulse rates for patients PO6 and PO2. Seven electric signals differing in pulse rate were presented in random order in a single interval task. The 7 stimuli were assigned response numbers from 1 to 7 in order of increasing pulse rate. The patients were instructed to identify each signal presentation as one of the 7 pulse rates, and to respond by nominating the appropriate response number. The pattern of response is presented as histograms representing the percentage of times each response number was nominated for a particular pulse rate. Perceptual sensitivity index, d′, between two successive pulse rates was computed on the basis of these histograms [137].

In another series of experiments we studied the ability of patients to identify different pulse rates after the loudnesses of the stimuli were also balanced. The results for two patients are shown in figure 76. These are confusion matrices in the form of response histograms. Seven stimuli were presented at different rates, and the subjects were asked to identify the stimulus. As can be seen, accurate identification only occurred up to a rate of 159 pulses/s for PO6 after which the histograms smeared. This was confirmed by the d′ values, which were calculated on the overlap between two successive histograms. The results for PO2 showed the ability to identify changes in rate of stimulation up to 501 pulses/s. The results in the six patients studied showed that the maximum rate that could be identified varied from 150 to 600 pulses/s. This is illustrated in figures 77 and 78.

It is not clear how the limited identification of pulse rate can be explained by the firing patterns of auditory nerve fibres. However, in the physiological studies, the failure of electrical stimulation to produce a

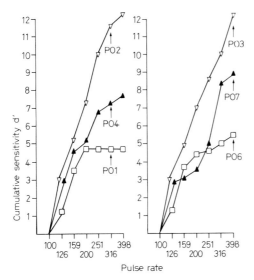

Fig. 77. Cumulative d' curves for the identification of 7 pulse rates in the lower range of the 6 patients. The 6 d' measures are plotted in cumulative form as a form of pulse rate for each patient. Some of the symbols are displayed sizeways for clarity [137].

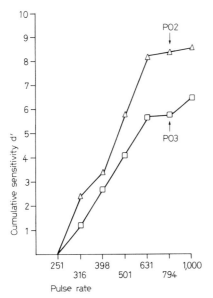

Fig. 78. Cumulative d' curves for the higher pulse-rate range for patients PO2 and PO3. PO2 and PO3 were the only two patients who felt confident in performing the identification task at these higher pulse rates [137].

more stochastic firing pattern and failure to induce a series of interpulse intervals could be key factors. It is also of interest that these psychophysical results for rate of stimulation were similar to those obtained some years earlier for behaviourally conditioned cats [29, 30, 141].

The results from the studies described above showed that the discrimination of pulse rate in cochlear implant patients was not possible above 600 pulses/s which is too low for the perception of important speech frequencies, especially the second formant frequencies which are in the range 850–3,000 Hz and which convey most information for speech intelligibility.

Place of Stimulation

As there were serious limitations on the coding of speech frequencies by rate of stimulation we carried out a series of investigations on the coding of frequency by place of stimulation. Studies on the identification of electrode position were carried out in a number of ways. In one study variations in pitch or sharpness with electrode position were evaluated using a paired comparison procedure. Patients were asked to say whether the sensation at one electrode was sharper or duller than the sensation at another.

The results for our first two patients using ten channels of pseudobipolar stimulation are shown in figure 79 [134]. Each 'standard' electrode was paired with itself and the other 'comparison' electrodes, and the patient asked to indicate whether the 'comparison' was duller or sharper than the 'standard'. The patients also equated sharpness with a high pitch and dullness with a low pitch. As can be seen there is little scatter of results across the diagonal indicating good place discrimination. The electrodes were also ranked from sharp to dull or high to low in an orderly fashion along the cochlea consistent with the place theory of pitch perception. The exception was, however, electrode 9 which was in the region of the round window, and the low or dull percept could have been due to the extracochlear spread of current from the site.

The psychophysical results for pitch perception as a function of electrode place were also recorded subsequently for the 22 electrode bipolar stimulus system used by the Cochlear Pty Limited receiver-stimulator. A sharpness or pitch-ranking experiment was also carried out. The results are shown in figure 80. The electrodes are numbered from 1 to 22 in a basal to apical direction. The sensation at the first electrode was compared with the sensation at a second electrode. Three presentations were

Standard stimulus (electrode No.)

Comparison stimulus (electrode No.)	9	1	2	3	4	6	7	8
9	D	D	D	D	D	D	D	D
1	S	D	D	D	D	D	D	D
2	S	S	D	S	S	D	D	D
3	S	S	S	D	D	D	D	D
4	S	S	S	S	D	S	D	D
6	S	S	S	D	S	D	D	D
7	S	S	S	S	S	S	D	D
8	S	S	S	S	S	S	S	D

Patient MC-1

Second electrode

First electrode	1	3	6	9	12	14	16	18	20
1	–	0	0	0	0	0	0	0	0
3	3	–	0	0	0	0	0	0	0
6	3	3	–	1	0	1	0	0	0
9	3	3	3	–	2	0	0	0	0
12	3	3	3	2	–	0	1	0	0
14	3	3	3	3	2	–	0	0	0
16	3	3	3	3	3	3	–	0	0
18	3	3	3	3	3	3	3	–	0
20	3	3	3	3	3	3	3	1	–

Fig. 79. Sharp-dull ranking data matrix for patient MC1. The order of the columns in the data matrix gives an approximate ordering of electrodes with respect to sharpness. The electrodes are numbered from 0 to 9 in the basal direction. A fixed pulse rate of 100 pulses/s was used. Each standard electrode was paired once with itself and the other 7 comparison electrodes and the patient was to identify whether the comparison was duller or sharper than the standard. Each cell of the matrix contains an S if the comparison was sharper than the standard, a D is the comparison was the same or duller than the standard. The same responses are indicated by the underlined Ds. The ordering of the columns (or rows) was adjusted so that the data matrix was the closest fit to a triangular pattern. The dull sensation on electrode 9 in the vicinity of the round window is likely to be the result of an extracochlear current path [134].

Fig. 80. Sharpness ranking response matrix. Electrode ranking data matrix for a recently implanted multiple-channel cochlear prosthesis patient. Each cell of the matrix represents a certain electrode pair in a particular order. The number in each cell corresponds to the number of times a second pair was identified by the patient as being sharper than the first. Each electrode pair was presented three times and thus a matrix entry of three indicates that the second electrode was chosen as sharper every time. A zero entry indicates that the first electrode was chosen as sharper every time. The electrode numbers correspond to positions in the cochlear numbering from the basal end [65].

made. If the first electrode was lower in pitch or duller it was scored as a 1 or if it was higher it was 0. For perfect place pitch recognition there should be 3 on one side of the diagonal, and 0 on the other side. As can be seen, pitch discrimination on a place basis is very good, although not perfect.

Another important psychophysical finding from these studies was that patients could give vowel labels for steady-state stimuli at different electrodes or, alternatively, they could learn to recognize different vowels when different electrodes were stimulated. The second formant frequency

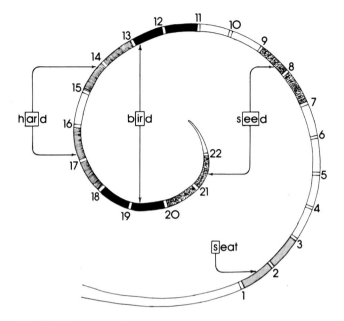

Fig. 81. Diagram showing the electric patterns of stimulation in the cochlea for the first and second formants of some vowels and consonants.

of the vowel was the frequency of best response for the particular electrode site. This is illustrated in figure 81. The vowel /i/ as in 'seed' has the highest second formant frequency and was perceived when the most basal or highest frequency electrode was stimulated, and is at a location normally most sensitive to 2,300–2,800 Hz; on the other hand, the vowel /3/ as in the word 'bird' has a low second formant frequency and was perceived when an apical or low-frequency electrode was stimulated, which is at a location normally most sensitive to 1,400 Hz. The second formant of the vowels in between also corresponded to the site of stimulation or frequency location especially after some initial training.

Combined Rate and Place of Stimulation

Having examined the perception of pitch as a function of rate and place of stimulation, it was considered important to also determine whether patients could in fact perceive two different and separate per-

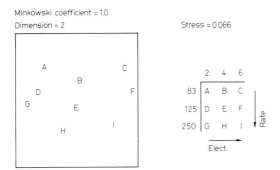

Fig. 82. Two-dimensional configuration representing the dissimilarities amongst the 9 electrical stimuli which were the nine possible combinations of 3 electrode positions and 3 repetition rates, obtained by multidimensional scaling in the left half of the stimulus matrix showing correspondence between the alphabetical symbols and the electrode and repetition rate arrangement in the right half [135].

cepts for rate and place of stimulation when both parameters were presented together. The results of this study were analysed by multidimensional scaling, and are shown in figure 82. From this it can be seen that the 2-dimensional solution of the results, shown on the left, is a close approximation to the stimuli shown on the right. This is confirmed by a low-stress value.

These findings indicate that it should be possible to code two separate pitch sensations in speech at the one time. For example, the low pitches of voicing could be perceived as rate of stimulation, and the higher pitches of the second formants as place of stimulation.

Time Varying Component Pitch Sensations

The psychophysical studies just described were for steady-state stimuli that had a duration of at least 200–300 ms, which is the duration of a vowel. Speech is, however, a dynamic stimulus. There is a slowly varying fundamental or voicing frequency and consonants, in particular, have frequencies that change rapidly over a duration of about 20 ms. For this reason the psychophysics of stimuli where either the place or rate of stimulation was varied over time, were studied. The results are shown in figures 83 and 84. The results for varying the place of stimulation over a short duration are shown in figure 83. The percentage judgments called

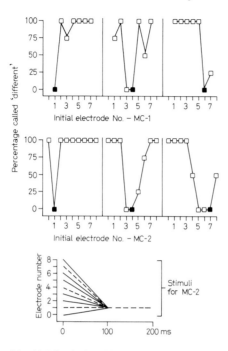

Fig. 83. Discriminability of time-varying electrode position with final electrode position at electrode 1 (stimulus set A). The bottom plot shows the stimuli used in this study
for patient MC-2. Over the first 100 ms of each stimulus the electrode position varied in an
orderly fashion from the variable initial electrode position to a fixed final electrode position (electrode 1), which persisted for the latter 100 ms of the stimulus. A constant pulse
rate of 100 pulses/s was used. The middle plot shows the result for MC-2. The stimuli with
initial electrode positions at electrodes 1, 4 and 7 were used as the standard stimuli in the
three traces respectively in the middle and were indicated by the dashed trajectories in the
bottom plot. For each trace in the middle plot, the standard 'stimulus' (solid square) was
paired 30 times with itself, and four times with each of the eight 'comparison' stimuli,
whose initial electrode positions were specified on the abscessa [134].

different were plotted against the initial electrode of a pair stimulated
separately with variable time intervals between the two. When the stimuli
went from electrode 1 to 1 there was no difference. When it went from all
other electrodes 2, 3, 4 to electrode 1 they were described as different
100% of the time regardless of durations of 25, 50 and 100 ms. This ability of patients to reliably detect a shift in electrode position over a duration of 20 ms indicates that place of stimulation is appropriate to code the
short frequency transitions of the second formants in consonants.

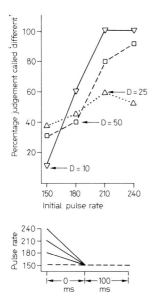

Fig. 84. Discriminability of time varying pulse rate for three durations (D, S): 25, 50, and 100 ms. A separate study was conducted for each D. The lower half shows the stimuli used in each study. Over the first Dms of each stimulus, the pulse rate varied linearly from a variable initial pulse rate to a fixed final pulse rate of 100 pulses/s, which persisted for the latter 100 ms of the stimulus. The stimuli were on electrode 1 throughout the (D+100)ms. The upper half shows the results as the percentage of the judgments called 'different' as a function of the initial pulse rate. The initial duration, D, appears as a parameter. A same-different procedure was used, and the percentage 'different' results at the 'standard' (solid symbols) were false alarms. Note the degradation in discrimination performance for an increase in initial duration, D. However, it should be noted that the discrimination performance for the time-varying electrode position was independent of the duration of the signal [134].

In this study the pulse rates were varied down to a steady 150 pulses/s over durations of 25, 50 and 100 ms (fig. 84). The percentage judgments called different were plotted for pulse rates which varied from 240 to 150 pulses/s for durations of 25, 50 and 100 ms. As can be seen there was a marked degradation in performance from 100 to 25 ms. This indicates that pulse rate is not adequate to convey the fast frequency changes that occur in consonants, but is appropriate for the slower frequency charges for voicing.

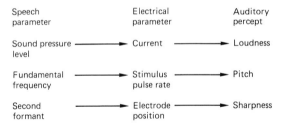

Speech parameter		Electrical parameter		Auditory percept
Sound pressure level	⟶	Current	⟶	Loudness
Fundamental frequency	⟶	Stimulus pulse rate	⟶	Pitch
Second formant	⟶	Electrode position	⟶	Sharpness

Fig. 84. Summery of the speech-processing strategy.

A Speech-Processing Strategy Converting Voicing to Rate of Stimulation and Second Formant to Place of Stimulation

As a result of the psychophysical studies a speech processing strategy for cochlear implant patients was developed and substantiated. A speech processing strategy was developed that would extract speech features such as formants as these are important cues in intelligibility, and their extraction and presentation could make speech understanding easier for the patient. In particular, the speech processing strategy extracted the fundamental or voicing frequency, the second formant frequency, and the amplitude of the speech waveform. The frequency of the second formant determined the site of stimulation. The rate of stimulation depended on the voicing frequency. However, with unvoiced speech a random pulse train was produced, and the current amplitude was related to the amplitude of the speech waveform. This is summarized in figure 85. Figure 86 shows an example of how the speech processing strategy was actually implemented. The first, second and third formant frequencies are shown for the word wit or its phonetic transcript /w//I//t/. As can be seen the second formant rises over a short time interval for the consonant /w/, remain steady for long period over the duration of the vowel /I/ and there is then a burst of high-frequency energy for the unvoiced phoneme /t/. The electrical stimuli produced by the cochlear implant in response to this word are represented by the vertical lines. For the /w/ sound the place of stimulation shifts from a low- to high-frequency area of the cochlea, and

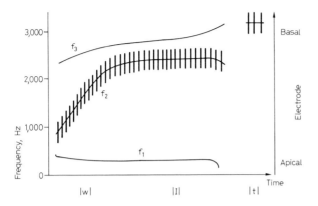

Fig. 86. Implementation of the speech-processing strategy as seen for the word wit or its phonetic transcript /w/ /I/ /t/. The vertical bars represent the electrical pulses.

as it is a voiced sound the spacing between the pulses is proportional to the voicing frequency. The rate of stimulation remains steady for the vowel and again the period of stimulation is proportional to the voicing frequency. Finally, for the /t/ sound which is a short burst of high frequency noise, an electrode in the basal area is excited. As it is unvoiced a low randomized pulse rate is used as this was identified by the patient as noise-like. Speech perception studies reported below established that this speech processing strategy converting the fundamental frequency to rate of stimulation, second formant to place of stimulation and speech amplitude to current level helped patients understand running speech when using electrical stimulation alone and especially when combined with lip-reading.

Two-Component Pitch Sensations. Two Sites of Stimulation

Further psychophysical studies were undertaken to determine how to increase the amount of information received by patients in order to develop improved speech-processing strategies.

Our first hypothesis was that the addition of first formant information presented on a place basis or as site of stimulation would improve speech-processing performance.

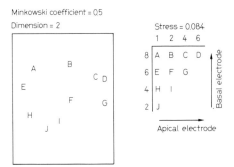

Fig. 87. Two-dimensional configuration representing the dissimilarities among the 2-electrode stimuli: the spatial configuration obtained by multidimensional scaling in the left half and the stimulus matrix showing correspondence between the alphabetical symbols and electrode pairs in the right half [136].

To help establish that this was possible we carried out a psychophysical study to see if patients could actually perceive a two component sensation when two pairs of electrodes were activated. Again a multidimensional scaling technique was used, and the results are shown in figure 87.

A 2-dimensional configuration provided the best approximation for the data. The results, therefore, showed that a speech-processing strategy which converts acoustic first and second formants to electrical stimulation at two separate sites should be possible.

Speech-Processing Strategy Converting Voicing to Rate of Stimulation and First and Second Formants to Rates of Stimulation

The speech processing strategy was implemented and evaluated on normal hearing subjects using an acoustic model of electrical stimulation [13–15], and also on cochlear implant patients. In both cases the addition of first formant information on a place basis was shown to result in better speech perception than when using a processor which provided voicing as rate of stimulation and second formant as place of stimulation. The results are discussed below.

Speech Perception in Postlingually Deaf Adults

The overall aim of our research has been to develop a cochlear implant that will help profoundly-totally deaf patients understand speech, and also assist them in recognizing environmental sounds. The studies are summarized in table 14. It was considered that high fidelity sound would not be possible with an implant due to limitations in reproducing the physiological responses in a population of nerve fibres. It was considered the most important task would be to help patients understand running speech, and that if they could do this they would also have an ability to recognize environmental sounds. As communication is such an important part of our lives, and as this is often severely affected in the case of the profoundly-totally deaf, it was considered that improved speech understanding should be the main thrust of the research we commenced in 1967.

To maximize the amount of information transmitted to patients, and give them the best chance of understanding speech, it was considered desirable to develop a prototype receiver-stimulator that provided multiple-channel stimulation using multiple-electrodes. This decision was based on physiological, psychophysical and speech research data. Firstly, auditory physiological research had established that both place and timing were important in coding frequency. Our initial unit response and field potential data [26] also showed that with electrical stimulation there would probably be a strict upper limit on the rate of stimulation that could be perceived as pitch on a single channel. Secondly, psychophysical studies [143] had shown that perceived pitch depended on the site of cochlear stimulation. Our psychophysical studies on cats [29, 30, 141] showed there was a limit of 600–800 pulses/s above which changes in stimulus rate could not be perceived, and this supported the idea that multiple-electrode stimulation should be carried out to maximize the perception of speech. Thirdly, speech research [84] had shown that at least ten frequency bands were necessary for the satisfactory transmission and perception of speech.

Table 14. Speech research studies

---

1 Speech processing strategy – physiologically based
2 Speech processing strategy – feature extraction F0-F2
3 Speech perception results for the speech processor extracting the fundamental (F0)
   and second formant (F2) frequencies
   A Laboratory-based speech processor
      Open-set word and sentence tests
      Speech tracking
      Vowels and consonants
   B Prototype portable speech processor
   C Analysis of speech information transmitted
   D Wearable speech processor for clinical trial
4 Speech-processing strategy – feature extraction F0-F1-F2
   A F1 (rate) – F2
   B F0-F1-F2
      Psychophysics
      Acoustic model
      (a) Psychophysics
      (b) Speech perception
      (c) Comparison of speech processing strategies
5 Speech perception results for speech processor extracting the fundamental (F0), first
   formant (F1) and second formant (F2) frequencies
   A Comparison of F0-F2 and F0-F1-F2 strategies
      Open-set word and sentence tests
      Speech tracking
      Vowels and consonants
      Analyses of speech information transmitted
      Performance in quiet and in noise
6 Speech perception results in languages other than English

To achieve these goals multiple-electrode, multiple-channel stimulation would be required, and this meant an implantable unit or percutaneous plug. As our research on animals had shown that percutaneous plugs were associated with chronic infection it was considered in patients' interests to develop a fully implantable prototype receiver-stimulator unit that would allow psychophysical studies to be undertaken, and speech-processing strategies to be evaluated.

Our prototype receiver-stimulator was first implanted on 1st August 1978 in a totally deaf patient who had lost all detectable hearing following a head injury 18 months previously. This patient then underwent a series of psychophysical tests to determine the percepts obtained for different stimulus parameters.

It has always been a fundamental part of our speech perception research to learn what basic information is transmitted to the patient in order to develop satisfactory speech-processing strategies. The psychophysical studies are discussed above. They confirmed our animal behavioural studies, showing that there were severe limitations on the perception of rate of stimulation in conveying pitch. This would prevent single-channel stimulation resulting in an optimal speech processor.

Speech-Processing Strategy – Physiologically Based

While studies were being undertaken to determine the psychophysics of electrical stimulation a speech processor that encoded sound on physiological principles was evaluated on the patient (fig. 88). This speech processor was designed to model basilar membrane motion, and the fine timing of auditory nerve firing [93]. This speech processor was tested on the first patient, and speech perception was not satisfactory. This was thought to be due in part to the fact it was providing simultaneous stimulation at different electrodes. The electric fields produced by simultaneous stimulation were summing, and leading to unpredictable variations in loudness. These results showed that one could not simply take a speech signal and hope to present it as a series of electrical stimuli without some form of pre-processing, or an understanding of the percepts obtained, or a knowledge of the spread of current within the cochlea.

Speech-Processing Strategy – Feature Extraction

For the above reasons it was considered desirable to stimulate electrodes non-simultaneously. That is, each different electrode was excited with a very short time interval between the stimuli, so their electric fields did not interact [51]. Secondly, it was considered important to carry out some form of pre-processing of the speech signal, and formant extraction was used. It was appropriate to use a formant-based speech processor as speech research studies had shown the importance of formants as cues in the understanding of speech [56, 74, 92, 127]. This is illustrated in

Fig. 88. Photograph of a real-time benchtop speech processor used to present the spectral frequencies of speech as a physiologically based code [93].

figure 89. Here we can see that the sound energy is concentrated in frequency bands called formants. These formants are essential in distinguishing vowels, and also for consonants. For example, in figure 90 we see how the first and second formants vary in frequency for the Australian vowels /i, I, ε, ae, u, ɜ, ʌ, a, D, ʊ, ɔ/ [6].

As a result of the psychophysical findings discussed in the chapter above a speech-processing strategy was developed in which the voicing or fundamental frequency was extracted using a 400-Hz low pass filter and used to stimulate electrodes at a rate proportional to the voicing frequency. It was considered appropriate to code voicing as rate of stimulation because the average voicing frequency for males is 120 Hz, and

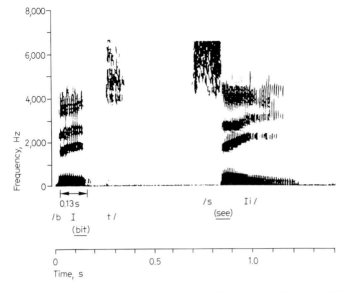

Fig. 89. The frequency spectra showing the formants for the words 'bit' and 'see'.

females 225 Hz. The frequency varies slowly in frequency when conveying intonation. As the psychophysical studies showed that rate of stimulation in the range required for voicing could be perceived by the patient this was an appropriate coding strategy. Unvoiced sounds were considered to be present if the energy of the voicing frequency was low in comparison to the energy of the second formant frequency, and coded by using a low random pulse rate, as this was described as rough and noise-like in quality.

The second formant frequency was estimated using a zero-crossing detector at the output of a 750-Hz high-pass filter, and an appropriate electrode stimulated on a place basis. The electrode stimulated was determined from pitch scaling experiments. The electrodes towards the basal end of the implanted cochlea were found to produce a high-pitch or sharp sound, while electrodes further into the cochlea produced lower-pitched sounds. Thus second formant frequencies were assigned to electrodes in an orderly fashion along the length of the implanted cochlea.

The amplitude of the sound energy of the second formant was used to set the current level of the stimulating electrode. This current level was set within the dynamic range recorded for each stimulus electrode.

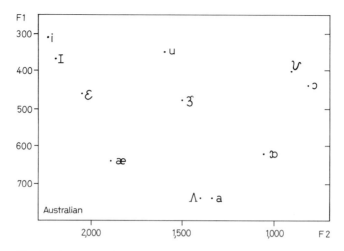

Fig. 90. The average first and second formant frequencies for the Australian vowels /i, I, ɛ, ae, u, 3, ɓ, a, b, ʋ, ʏ [6].

The coding strategy was implemented on the laboratory-based computer shown in figure 48, and is summarized in figure 85. The frequencies and amplitudes were sampled every 20 ms, and an example of how the word wit (transcribed phonetically as /w/, /I/, /t/) was coded, can be seen in figure 86. Figure 86 shows the acoustic analysis of the word wit into formants with the vertical bars representing the electric pulses. The phoneme /w/ has a second formant which rises, and for this reason electric pulses were presented to different electrodes where positions along the cochlea correlated with the changing second formant frequency. /w/ is also a voiced phoneme therefore the rate of stimulation at each electrode was proportional to the voicing frequency. The phoneme /I/ has a steady second formant frequency of approximately 2,000 Hz, and consequently an electrode at a site in the cochlea which would normally respond to this sound frequency was excited. As /I/ is voiced the stimulation rate was also proportional to the voicing frequency. Finally, the phoneme /t/ has a high second formant, and consequently a basal electrode was excited. Furthermore, it is an unvoiced sound, and this resulted in a stimulus at a low and random rate as this was perceived as noise-like by the patient.

It is important to note that this speech-processing strategy partially simulated the physiological situation for speech sounds. With speech, neural firing in response to voicing occurs along the length of the cochlea, but is broadly tuned at the apical end. Furthermore, the second formant frequency results in a localized neural excitation which is more finely tuned in the region where the frequency of best response is the same as that of the second formant frequency [91, 118]. With the cochlear implant speech-processing strategy, the second formant excited a similar localized area of the cochlea at a rate that was proportional to the voicing frequency.

The main differences from the physiological situation were, firstly, that only one site was stimulated at a time and, secondly, that all nerve fibres in a stimulated group fire together or deterministically, and not randomly or stochastically as occurs normally.

Speech Perception Results for the Speech Processor Extracting the Fundamental and Second Formant Frequencies

Laboratory-Based Speech Processor

Having developed the feature extraction or formant-based speech processor on the laboratory computer the next step was to evaluate it on our first patient to see if it helped him in understanding speech. At the time it was not clear what tests would be the most appropriate to measure improvements in speech understanding when using the cochlear implant. It was considered, however, the best way to do this was to carry out a number of standard audiological tests that were in use to assess deaf patients with hearing aids. In particular, open-sets of phonetically balanced monosyllabic words, and open-sets of CID sentences were selected as these were well-established tests, and the results were shown to correlate with a patient's hearing disability. It was also appreciated that it was very important to standardize the tests in all ways so that one could be objective about the cochlear implant performance.

The tests were standardised by firstly making sure that they were open-sets. This meant that they must have been taken from the standardized lists of words or sentences, and that they should not have been previously presented to the patient. Secondly, it was necessary to indicate whether the material was presented live-voice or pre-recorded on video tape. Thirdly, it was important to indicate whether the test material was presented once only or with repetitions. Fourthly, it was essential to specify whether a hand-held or ear level microphone was used, and the distance of the microphone from the loudspeaker. Finally, it was important to define the listening conditions, for example, whether the test was carried out in a noisy or sound-treated room. Ideally, it is desirable to know the acoustic properties of the recording room and the residual sound levels.

Open-Set Word and Sentence Tests

These standardized word and sentence tests were carried out on our first patient [39, 40] and they showed that he was able to understand some open-set speech with electrical stimulation alone, and that there were marked improvements in scores when using electrical stimulation combined with lip-reading compared to lip-reading alone. When using open-sets of monosyllabic AB words [17] he obtained a 10% score for electrical stimulation alone, and there was a 300% improvement in score when using electrical stimulation combined with lip-reading, compared to lip-reading alone. In this case the score went from 10% for lip-reading alone to 40% for the combined condition. In the case of the CID sentence test he obtained an open-set score of 14% for electrical stimulation alone. With electrical stimulation combined with lip-reading the score was 68% compared to 14% for lip-reading alone, which was a 386% increase.

Having established that the formant-based, multiple-electrode, multiple-channel cochlear implant could help one postlingually deaf adult patient understand running speech, it became important to know whether the speech-processing strategy could be generalized and help other patients. Another important question was over how many years could a postlingually deaf patient retain his memory and central processing ability for speech sounds. For these reasons a second patient had a cochlear implant operation on 17 July 1979. This patient was a 64-year-old man who had a progressive loss of hearing since 28 years of age, and had been profoundly-totally deaf for 13 years prior to surgery. It was very encouraging for the programme that when the speech processor was used at his first test session he was able to understand speech when used in conjunction with lip-reading. The open-set word and sentence test results for the first (MC-1) and second (MC-2) patients are shown in table 15 [39, 40]. These results helped support the view that the speech processing strategy would apply to postlingually deaf patients in general, and that patients would receive benefit even though they had been deaf for many years. This hypothesis of course needed to be tested further on more patients.

Table 15. Opcn-scts of AB words and CID sentences: averaged results for patients MC-1 and MC-2 [39, 40]

	AB words, %		CID sentences, %
	phoneme	word	
EA	20	5	11
LA	58	20	24
EL	76	50	83

EA = Electrical stimulation alone; LA = lip-reading alone; EL = electrical stimulation and lip-reading.

Speech Tracking

As it was important to establish that the speech-processing strategy could help patients understand running speech it was considered desirable to use the speech tracking test [62] as a more direct measure of a patient's ability to understand running speech. The test requires the patient to repeat back sentences or phrases read from a book by the tester. When a sentence is not repeated correctly, it is presented again, or broken into smaller sections until the patient has correctly repeated it. The test score is the number of words correctly repeated by the patient divided by the time taken. It was considered that if the speech tracking test were carried out in a standardized way it would be reasonably objective, and would be a good measure of a patient's ability to understand running speech. The results of the tracking test [97] on patients MC-1 and MC-2 are shown in table 16.

Patient MC-1 improved in tracking rate from 5.5 words/min for lip-reading alone to 34.0 words/min for lip-reading combined with electrical stimulation (518% increase). Furthermore, in the combined condition he reached a score that was 32% of a normal hearing subject. Similar good results were obtained for the second patient MC-2. His tracking score went from 12.9 words/min for lip-reading alone to 40.4 words/min for electrical stimulation combined with lip-reading (213% increase). In the combined condition he achieved a score that was 38% of that for a normal hearing subject.

Table 16. [97]

	MC-1	MC-2
Average tracking rates (words per minute)		
EL	34.0	40.4
LA	5.5	12.9
Percentage of performance of normal hearing subjects		
EL	32	38
LA	5	11

EL = Electrical stimulation plus lip-reading; LA = lip-reading alone.

Vowels and Consonants

Tests were also performed to measure vowel and consonant recognition for electrical stimulation alone using the laboratory-based speech processor. The average results for a closed-set of six vowels /i, a, ɔ, I, ʌ, δ/ in an H-vowel-D content (heed, hard, hoard, hid, hud, hod) were an average of 77% correct for the two patients [41, 131, 132]. The average results for a set of the 10 consonants /k, p, s, ʃ, t ʃ, f, w, j, r, n/, in a consonant-vowel context were 35% [131, 132], for a set of the eight consonants /p, t, k, b, d, g, m, n/ in a vowel-consonant-vowel (VCV) context 37% [39, 40], and set of 12 consonants /p, t, k, b, d, g, m, n, f, v, s, z/ in a vowel-consonant-vowel context 36% [131].

Prototype Portable Speech Processor

Having shown that the laboratory computer-based speech processor would enable the two postlingually deaf adult patients to understand speech, the next task was to implement the speech processing strategy in a portable unit. This was achieved in the Department of Otolaryngology in 1980, and the unit which measured $15 \times 15 \times 6.5$ cm is shown in figure 49.

It was then necessary to determine whether the portable speech processor could perform as well as the laboratory computer-based processor. This was done by presenting open-sets of AB words and CID sentences as discussed above. The results from patient MC-1 for both speech processors are shown in table 17. From the results [133] it can be seen that the performances were similar indicating that the wearable speech processor had been well engineered to reproduce the capabilities of the laboratory computer-based processor.

Table 17. A comparison of results in patient MC-1 for the laboratory-based and prototype portable speech processor [133]

	Laboratory-based[a] speech processor			Prototype portable[b] speech processor		
	EA	LA	EL	EA	LA	EL
AB words	10	10	40	0	30	60
AB phonemes	20	53	73	23	70	83
CID sentences	14	14	68	10	38	76

[a] Tests pre-recorded.
[b] Tests live voice.
EA = Electrical stimulation alone; LA = lip-reading alone; EL = electrical stimulation with lip-reading.

Analysis of Speech information Transmitted

Having shown that the formant-based speech-processing strategy used with the multiple-electrode, multiple-channel implant could help patients understand running speech it also became important to determine how it helped. Did the speech processor in fact perform on a multiple-electrode, multiple-channel basis, and were the results better than single-channel systems currently available?

It was considered the best method of obtaining this data was to test the patient with closed sets of nonsense syllables, and analyse the consonant confusions to determine whether voicing, manner or place information was transmitted to the patient.

In the first study [39] nonsense syllables were constructed from the consonants /b, p, m, d, t, n, g, k/ and were presented in a VCV framework using the vowel /a/ as in father. The results (table 18) showed that voicing, manner and place information was transmitted for electrical stimulation alone, and the results were also better when electrical stimulation was combined with lip-reading, compared to lip-reading alone. These results were consistent with the fact that the device was transmitting information for both timing (rate) as well as site of stimulation, and that this information was being used by the patient to recognize speech sounds. Voicing was coded by the speech processor as timing (rate) of stimulation, and used to make voicing distinctions, for example between /b/ and /p/. This

Table 18. Voicing, manner and place distinctions [39]

Patient	Test voice		EA	LA	EL
Laboratory computer-based speech processor /b, p, m, d, t, n, g, k/, percentage correct					
MC-1	F	voicing	94	62	98
		manner	75	60	73
		place	62	78	89
	M	voicing	83	59	86
		manner	66	63	76
		place	51	71	88
MC-2	F	voicing	86	65	96
		manner	71	64	77
		place	54	84	93
	M	voicing	71	54	68
		manner	76	66	66
		place	53	72	84
Portable prototype speech processor, percentage correct					
MC-1	F	voicing	94	64	98
		manner	83	66	93
		place	71	80	94
	M	voicing	82	62	90
		manner	86	73	85
		place	62	80	85

EA = Electrical stimulation alone; LA = lip-reading alone; EL = electrical stimulation plus lip-reading.

is important as these two sounds look the same on the lips, and the only difference is that /b/ is voiced and /p/ is not.

It was essential to know whether voicing was transmitted, as the first requirement of a cochlear implant is that it help speech reading. The fact that voicing information was transmitted was confirmed by the fact that the scores for voicing for electrical stimulation plus lip-reading were much better than those for lip-reading alone, and that the scores for electrical stimulation alone were similar to those for electrical stimulation plus lip-reading.

Manner and place information is also important in helping patients with lip-reading. This information is, however, especially important if a patient is to understand running speech without lip-reading help. In both cases the additional second formant information obtained by stimulating the cochlea on a place basis was helpful in providing these cues. Manner distinctions, for example the difference between the plosive /b/ and the nasal /m/, are perceived on the basis of the duration of the sound as well as other cues such as the first and second formant. Certain manner differences cannot be well read on the lips so it is useful to have additional information provided by the implant. The results do in fact show that the implant provided additional information to help with manner cues.

Information on the place of articulation is especially important for cochlear prostheses that are going to help patients understand running speech without lip-reading. For example the place of articulation is different for /p/ and /t/, but as they look quite different on the lips, place of articulation is not so essential for lip-reading help. Place of articulation is perceived by a number of cues, in particular the first and second formants which are in turn dependent on the site of stimulation in the cochlea. The results in table 18 show that information on place of articulation is provided by electrical stimulation alone, and also there is a small improvement in the scores for electrical stimulation plus lip-reading compared to lip-reading alone.

A more detailed consonant confusion study with 12 consonants [63] was undertaken on MC-1 using the portable speech processor. The information transmitted was determined for voicing, nasality, affrication, duration, and place, for lip-reading alone, electrical stimulation alone and electrical stimulation combined with lip-reading.

The information transmitted was calculated from confusion matrices. An example of the confusion matrices from a patient for lip-reading alone and for electrical stimulation combined with lip-reading are shown in table 19. The information transmission T (x, y) from x (stimuli) to y (response) in bits per stimulus when H (x) is the information available in bits per stimulus, is calculated as follows:

$$T_{percent}(x, y) = T(x, y)/H(x) \times 100.$$

The results for the percentage information transmission for closed-sets of the 12 consonants /b, p, m, v, f, d, t, n, z, s, g, k/ confirmed our previous findings [39] that electrical stimulation alone provided help with all speech features, and that the electrical stimulation plus lip-reading results were better than those for lip-reading alone. The results confirmed that the speech-processing strategy was providing information both by rate and site of stimulation within the cochlea. The information transmitted for nasality, affrication, duration and place was a help to lip-reading

Table 19.

Condition — Female speaker

Lip-reading alone — Response (32% correct)

Stimulus	b	p	m	v	f	d	t	n	z	s	g	k
b	10	4	6									
p	13	4	3									
m	8	2	10									
v				6	14							
f				10	10							
d						6	1	7	2	2	2	
t						4		1	8	4	2	1
n						7	12				1	
z						4	2		9	5		
s						1		1	12	5	1	
g						6		11			3	
k						6	1	7			5	1

Condition — Male speaker

Lip-reading alone — Response (28% correct)

Stimulus	b	p	m	v	f	d	t	n	z	s	g	k
b	13	3	4									
p	12	5	3									
m	8	5	6	1								
v				8	12							
f				12	8							
d						5		5	5	2	2	1
t						7		3	8	2		
n						8	10				2	
z						4	1		7	5	2	1
s						6		1	10	2	1	
g						5		9	1		2	3
k						4	1	6			7	2

Lip-reading with electrical stimulation — Response (65% correct) — Female speaker

Stimulus	b	p	m	v	f	d	t	n	z	s	g	k
b	12		8									
p		20										
m	7	2	11									
v				11	9							
f				7	13							
d						4		6			10	
t							19				1	
n						2		14			4	
z						1			16	3		
s									9	11		
g						3		3			13	1
k								2			5	13

Lip-reading with electrical stimulation — Response (75% correct) — Male speaker

Stimulus	b	p	m	v	f	d	t	n	z	s	g	k
b	15		5									
p		20										
m	1		19									
v				14	6							
f				8	12							
d						10	1	3			5	1
t							16				1	3
n							5	12			2	1
z									16	4		
s									5	15		
g							2	4			13	1
k											2	18

as well as the transmission of voicing alone, and also had the potential to help patients understand speech without lip-reading assistance.

A further study was also undertaken to help determine whether the speech perception results were better for the F0-F2 method of multiple-channel stimulation compared to single-channel F0 stimulation. The two speech-processing strategies were compared on our first patient. He had been using a speech processor which extracted the voicing (F0) and second formant (F2) frequencies. Although this patient was not trained for equal periods of time with both strategies, as he had experienced bet-

Table 20. Comparison of consonant speech feature information transmission in F0 and F0-F2 speech processors [45]

Speakers	EA		LA		EL	
	F0 %	F0-F2 %	F0 %	F0-F2 %	F0 %	F0-F2 %
Voicing	26	25	0	0	28	46
Nasality	5	10	3	5	16	43
Affrication	11	28	28	37	52	100
Duration	10	80	11	22	33	100
Place	4	19	70	69	70	80
Overall	35	42	54	47	63	75

EA = Electrical stimulation alone; LA = lip-reading alone; EL = electrical stimulation plus lip-reading.

ter speech perception with the F0-F2 processor, it was considered that he had sufficient experience with the F0 strategy during continual use of the F0-F2 processor to make valid comparisons. The results of the comparison [45] are shown in table 20 for the consonants: /b, p, m, v, f, d, t, n, z, s, g, k/. The results show that the F0-F2 speech processor provided more information than the F0 processor for all speech features except voicing. It is also interesting to compare the performances of the speech processors using speech perception tests. The results are shown in table 21. It should be noted that when the data was analysed statistically by the method described by Thornton and Raffin [129] the F0-F2 speech processor was significantly better than the F0 processor for all tests when using electrical stimulation alone. It was also better for all tests except the modified rhyme test (MRT) when using electrical stimulation combined with lip-reading.

Wearable Speech Processor for Clinical Trial

Having established the value of the prototype portable F0-F2 speech processor in helping profoundly-totally deaf adult patients with a postlingual hearing loss understand speech, a smaller wearable speech processor and more reliable receiver-stimulator for clinical trial by the FDA was developed by Nucleus Limited. This wearable speech processor allowed the F0-F2 speech processing strategy to be used, but also had provision

Table 21. Comparison of speech perception for F0 and F0-F2 speech processors in MC-1 [45]

Test	Condition	F0 Speech processor	F0-F2 speech processor
16-SP			
Words	EA	28	100[a]
MRT words	EA	24	50[a]
	LA	62	58
	EL	76	84
AB words	EA	0	25[a]
	LA	10	20[a]
	EL	20	50[a]
AB			
Phonemes	EA	10	42[a]
	LA	53	53
	EL	53	72[a]
WIPI words	EA	0	28[a]
WIPI			
Phonemes	EA	8	50[a]
CID sentences	EA	0	22[a]
	LA	42	36
	EL	60	98[a]

NA = Not applicable.

[a] Significantly different from F0 speech processor score at 5% level [129].

F0 = Speech processor extracting fundamental or voicing frequency; F0-F2 = speech processor extracting fundamental and second formant frequencies; 16-SP = 16 spondee; MRT = modified rhyme test; AB = Arthur Boothroyd; WIPI = word intelligibility by picture identification; CID = Central Institute for the Deaf; EA = electrical stimulation alone; LA = lip-reading alone; EL = electrical stimulation plus lip-reading.

for other strategies to be assessed. The pre- and postoperative tests used to assess the patient are listed in table 11, and were adapted from the minimal auditory capabilities (MAC) battery [109].

An initial clinical trial of the receiver-stimulator and speech-processor manufactured by Nucleus Limited was first carried out at the Royal Victorian Eye & Ear Hospital on six patients in 1982 to help establish that the newly engineered device could perform at least as well as the prototype device developed by the University of Melbourne. Furthermore, it was also important to be sure that the initial encouraging results

Table 22. Open-set scores for electrical stimulation alone [65]

Test	Patient								Mean
	1	2	3	4	5	6	7	8	
Monosyllabic words (NU6, n = 50)	10	4	6	14	0	0	10	4	6
AB words (phoneme score) (30 words, n = 90)	40	30	60	60	20	23	48	37	40
CID sentences (2 tests) (n = 100 key words)	15	4	24	38	2	2	58	8	19

on the first two patients would apply to a larger population. The results for four, six, eight and ten patients [46, 64–66, 112] confirmed that they were able to obtain significant help in understanding speech when the speech processor was used in combination with lip-reading compared to lip-reading alone, and about 50% were able to understand some speech without lip-reading. The results for open-sets of monosyllabic words, open-sets of AB words scored phonetically, and open-sets of words in everyday CID sentences, for electrical stimulation alone, are shown in table 22. From this it can be seen that there was considerable variation in scores. For example, the open-set CID sentence scores for electrical stimulation alone varied from 2% to 58%. The mean scores for monosyllabic words were 6%, AB words 40% and CID sentences 19%.

If these results on the wearable speech processor for clinical trial on eight patients (table 22) are compared with those for the laboratory-based and prototype portable speech processors in table 17, it can be seen that the mean results were better for the wearable speech processor. For example, the scores for AB phonemes for electrical stimulation alone were 40% for the wearable speech processor, and 23% for the prototype portable speech processor, and for the CID sentences they were 19% for the wearable speech processor compared to 10% for the prototype portable speech processor. The difference was probably even more significant as the tests used with the prototype portable speech processor were delivered live-voice with a familiar speaker whereas the tests for the wearable speech processor were prerecorded with an unfamiliar speaker.

The average tracking rates for the wearable speech processor with eight patients was 36.2 words/min for electrical stimulation combined with lip-reading compared to 17.3 words/min for lip-reading alone. On the other hand the average score for the first two patients using the laboratory-based speech processor were 37.2 words/min for electrical stimulation plus lip-reading and 9.2 words/min for lip-reading alone. The results for other speech tests were similar for both the prototype portable speech processor and the wearable speech processor for clinical trial. For example the mean scores for consonants for electrical stimulation alone were 36% for the portable prototype speech processor and 43% for the wearable speech processor.

Not only did the wearable speech processor help the patients understand speech but also environmental sounds. With the environmental sound test in the MAC battery carried out without any practice, the average score was 27% correct.

As the initial clinical trial of the wearable speech processor and receiver-stimulator manufactured by Nucleus Limited gave speech results on eight patients that were at least as good as those obtained with the University of Melbourne's prototype device on two patients, a clinical trial for the FDA was extended to a number of centres in the USA, Canada and the FRG. These centres were initially: University of Iowa, Iowa; Baylor College of Medicine, Houston; Mason Clinic, Seattle; New York University, New York; Good Samaritan Hospital, Portland; University of Toronto, Toronto, Louisiana State University, New Orleans, and Medizinische Hochschule, Hannover.

The results on 40 patients from both Australian and North American centres have been analysed and are presented by Dowell et al. [67]. The results from other centres are comparable to those obtained in the initial clinical trial at the Royal Victorian Eye and Ear Hospital in Melbourne. The open-set CID sentence scores for 40 patients for electrical stimulation combined with lip-reading compared to lip-reading alone are shown in figure 91. The mean score for lip-reading alone was 51.9% (range 15–85) and for F0-F2 electrical stimulation plus lip-reading 87.0% (range 45–100). This was an average 68% improvement in scores for the combined condition. The 87% average score is not far from the 100% norm for the test, and indicates that the average patient should be able to understand considerable amounts of running speech when using F0-F2 electrical stimulation in combination with lip-reading.

It has been shown that CID sentence test scores may predict how well patients are able to comprehend running speech. Giolas et al. [83] were able to stimulate various degrees of deafness in normal hearing subjects

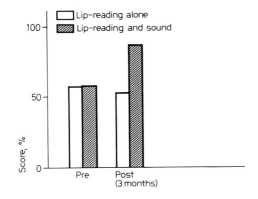

	Preop.		Postop.	
	LA	L and S	LA	L and S
Mean	55.7	56.7	51.9	87.0
SE	3.89	3.75	3.65	2.36
Range	14–89	18–100	15–85	45–100

Fig. 91. Mean CID sentence preoperative and postoperative test scores for F0-F2 electrical stimulation combined with lip-reading alone. Three months postoperative (n=40) [67].

by presenting filtered speech to them. The speech material was either CID everyday sentence lists, or Ulrich's samples of continuous discourse. In this way it was possible to relate sentence scores to the percentage of continuous discourse heard. From Giolas et al.'s findings it would appear that our average patient should have been able to perceive about 80% of connected discourse using electrical stimulation plus lip-reading. Furthermore, we have found that cochlear implant patient performances can improve considerably with time. The results reported above were recorded at the 3-month postoperative test session, and would have been very much better at 12 months.

The improvements possible over time are illustrated in figure 92. Preoperatively the mean open-set CID sentence score for hearing alone was 0.1% (range 0–4). The postoperative score for electrical stimulation alone at 3 months was 16.2% (range 0–58) and the score at 12 months was 39.7% (range 0–86). This was an average 145% increase in 9 months. One patient also obtained an 86% score for open-set CID sentences using electrical stimulation alone, and was close to the norm for this test.

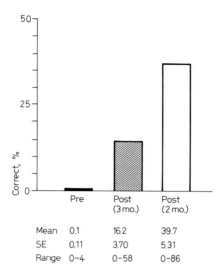

	Pre	Post (3 mo.)	Post (2 mo.)
Mean	0.1	16.2	39.7
SE	0.11	3.70	5.31
Range	0–4	0–58	0–86

Fig. 92. Mean CID sentence scores for hearing alone preoperatively, and F0-F2 electrical stimulation alone 3 months postoperatively and 12 months postoperatively (n = 23) [67].

The results for the open-set CID sentences indicate that a large number of patients should be able to understand running speech using F0-F2 electrical stimulation plus lip-reading. This was confirmed with the tracking results for the 40 patients which are presented in table 23.

From this it can be seen that the mean scores went from 19.2 to 48.6 words/min, a 138% increase. As the ceiling tracking rate for normal hearing subjects was 100 words/min it means the average patients were able to understand speech when using electrical stimulation plus lip-reading at approximately half the rate of normal hearing subjects. On the other hand, a few were able to track at near normal rates as indicated by the range going up to 100 words/min. Again the results with the F0-F2 wearable speech processor for clinical trial are similar to those for the F0-F2 prototype portable speech processor where a mean tracking score of 37.2 words/min was obtained for the two patients in the combined condition. The percentage improvement for these two patients was better, however, and averaged 366%.

Not only were patients receiving considerable help in understanding running speech when using electrical stimulation combined with lip-

Table 23. Speech tracking, words/min (n = 40) [67]

Score	Lip-reading alone	Electrical stimulation plus lip-reading
Mean	19.2	48.6
Range	6.38	25–100
SE	1.65	2–16

reading compared to lip-reading alone, but about a half reported they could understand significant amounts of running speech using electrical stimulation alone. This is consistent with the tracking results where 20 of the 40 patients were reported to be able to do the test using electrical stimulation alone. The results, however, were obtained over a number of test sessions for only 12 of the 20 patients. A mean tracking rate of 27.0 words/min was obtained and this was above the average obtained for lip-reading alone which was 19.5 words/min [21]. In fact nine of these 12 patients demonstrated a greater ability to communicate with electrical stimulation alone compared to lip-reading alone. This indicates the coch-lear implant was effective in a proportion of patients in providing speech understanding for electrical stimulation alone rather than acting only as a supplement to lip-reading.

This ability of some of the patients in understanding running speech using F0-F2 electrical stimulation alone was reported for a subgroup of 13 Melbourne patients [67] from the total of the first 40 who were combined from US, Canadian and Melbourne centres. In the subgroup, four of the 13 patients demonstrated the ability to understand connected speech without lip-reading help. These patients were able to repeat verbatim unknown material read by a tester at rates up to 35 words/min. They were also able to understand an average of 78% of key words in open-sets of CID everyday sentences using live voice and a hand-held microphone. They were also able to understand an average of 51% of key words in open-sets of everyday sentences presented over the telephone. These results confirm that for a proportion of postlingually deaf patients, the F0-F2 speech processor can not only act as an aid to lip-reading but restore effective speech understanding without lip-reading.

The ability of patients to use a telephone, especially for an interactive conversation, is very important in our society. This was first demonstrated on one of our patients [21]. Open-sets of CID everyday sentences were presented over the telephone, and the patient repeated an average of 21% of the key words correctly on the first presentation, and 47% when one repeat of the sentences was presented. In a group of three patients [51] the average open-set CID sentence test score over the telephone was 35% for the first presentation, and 50% when one repeat was allowed.

It was of interest to examine the vowel and consonant scores for the first 40 patients in the FDA clinical trial. The results were obtained from live presentation of closed sets of 11 vowels for the Australian and nine vowels for the North American patients. The vowels were in an h-vowel-d context. The consonants were /b, n, p, v, f, d, z, s, t, g, k/ for both groups and were in an /a/-consonant-/a/ context. The mean score for vowels was 55% (chance level 11%), and the score for consonants was 42.4% (chance level 8.5%). The consonant results showed a larger range of scores than for the vowels, with a few patients scoring appreciably better than the average, and a few finding the task very difficult. It is not clear why the patients who scored the highest on the consonant test were not necessarily those who scored the highest on the vowel test. An analysis of the results for the Australian and North American patients revealed no significant difference between the groups.

It was also of interest to compare the results for 12 consonants for the 40 patients using the F0-F2 wearable speech processor for clinical trial, and the small group of two patients who first had the F0-F2 laboratory-based and prototype portable speech processor. In the former case the mean score was 42.4% and in the latter case 36%. These results indicate that there was no real difference between the consonant data. These findings, therefore, help confirm that the wearable speech processor adequately realized the speech processing strategy first established on the initial group of patients in 1978 and 1979.

Speech-Processing Strategy – Feature Extraction
(First and Second Formants)

Although many patients were obtaining good results with the F0-F2 speech processor, they were not the same as those for normal hearing subjects. The challenge, therefore, was to develop speech processors that would peform better, and so help more patients communicate.

F1 (Rate) – F2

The first hypothesis to be tested was that an improvement would result if the speech processor presented first formant as well as second formant information. Firstly, a speech-processing strategy was developed in which rate of stimulation was used to code the first formant (F1), and place of stimulation was used for F2. In this speech-processing strategy, F1 stimulated whatever electrode was used to present F2. The coding strategy was similar to the F0-F2 one with the difference being that the stimulus was at the F1 rather than the F0 rate. F0 information was still transmitted because the periodic voicing energy used to excite the resonators in the mouth also led to groups of pulses at F0 intervals.

It was appropriate to use the coding strategy as our previous psychophysical studies described above had shown patients could detect differences in rate of stimulation up to 800 pulses/s, and this was partly within a normal F1 frequency range, which is from 300 Hz to 1,000 Hz. It did, however, depart from a physiological representation of speech as acute animal studies [91, 118] had shown that spike intervals at the F1 frequency only occur in neurons excited in a localized area by the F1 frequency. In the animal studies nerve spike intervals did not occur at the F1 frequency along the length of the cochlea as was the case with this particular F1-F2 strategy.

This F1 (rate) – F2 speech-processing strategy was evaluated on a cochlear implant patient whose psychophysics tests had shown that she had good rate discrimination up to 600–800 pulses/s. The results were as follows: vowels – F0/F2 – 41%, F1(rate)/F2 – 42%; consonants – F0/F2 – 26%, F1(rate)/F2 – 24%; speech tracking – F0/F2 – 32 words/min, F1(rate)/F2 – 34 words/min [98].

F0-F1-F2

The second example of a speech-processing strategy providing first and second formant information was one where both F1 and F2 were presented on a place basis, and F0 was used to stimulate each electrode along the cochlea at a rate proportional to the voicing frequency. As with the previously described F0-F2 strategy, no two electrodes were stimulated simultaneously, as this would have led to unpredictable variation in loudness. The F1 and F2 stimuli were always separated by a small interval in time no smaller than 0.8 ms.

This F0-F1-F2 speech-processing strategy was more physiologically based than the F1 (rate) – F2 strategy as the animal studies [91, 118]

Table 24. Relative pulse rate difference limens – acoustic model [14]

Subject	Pulse rate	Relative difference limens, %
A	100	3.1 ± 0.2
	200	7.9 ± 0.6
B	100	6.7 ± 0.7
	200	10.0 ± 0.9
C	100	2.2 ± 0.2
	200	4.2 ± 0.3

showed that the F1 and F2 frequencies stimulated neurones at appropriate sites along the cochlea, and that the low frequencies of voicing (F0) resulted in neurones along the cochlea firing at the same rate. The development of the speech processing strategy was also based on previous psychophysical studies, and an acoustic model of a cochlear implant used to evaluate speech-processing strategies on normally hearing subjects.

Psychophysics

Our psychophysical studies reported above [136] had shown that if two sites along the cochlea were stimulated near simultaneously (i.e. separated by a short interval in time) the patient could hear a two-component sensation. This, therefore, formed the psychophysical basis for presenting two speech formants by place of stimulation.

Acoustic Model

In addition, an acoustic model of electrical stimulation was developed and tested on normally hearing subjects [13, 14]. With the model a set of stimuli were generated from a pseudorandom white noise generator, the output of which was fed through seven separate bandpass filters corresponding to different electrode sites. The psychophysical tests showed similar results using the model on three normally hearing subjects compared to multichannel electrical stimulation on two cochlear implant patients. The psychophysical tests were: pulse rate difference limen measurements; pitch scaling for stimuli differing in pulse rate; pitch scaling and categorization of stimuli differing in filter frequency or electrode position; and similarity judgments of stimuli differing in pulse rate as well as filter frequency or electrode position.

Table 25. Relative pulse rate difference limens – electrical stimulation [14]

Patient	Pulse rate	Relative difference limen, %
MC-1 [134]	100	7
	200	5
MC-2 [134]	100	6
	200	8
MC-1 [14]	100	2.6 ± 0.4
	200	4.7 ± 0.8

Pulse rate difference limens were measured using a two-interval, forced-choice paradigm. The results are shown in table 24. If these are compared with the results for two cochlear implant patients (table 25) the agreement between the model and electrical stimulation for the test can be seen.

Pitch as a function of pulse rate was assessed using a two-interval ratio estimation procedure with a fixed standard. The pitch of stimuli with pulse rates from 50 to 1,000 pulses/s was investigated. The same procedure was used to obtain comparable cochlear implant data [13, 14]. The results are shown in figure 93. The shape of the sets of curves for the acoustic model and electrical stimulation are similar with a saturation of the pitch rates at about 300 pulses/s. This indicates that differentiating pitch above 300 pulses/s was not possible. The range of pitch ratios used by MC1 was quite different from the acoustic model subjects and no satisfactory explanation was found.

Pitch as a function of electrode position and filter frequency was evaluated by measuring pitch ratio after the loudness of the stimuli were matched. d′ distances between stimuli on successive filters and electrodes were calculated from the pitch ratios according to the method of Braida and Durlach [18]. The results are shown in figure 94. From this it can be seen that the results for the acoustic model filter or place of stimulation are similar to those obtained for different sites of electrical stimulation in the cochlear implant patients.

Finally, the ability of subjects and patients to judge stimuli differing in pulse rate as well as filter frequency or electrode position was evaluated using multidimensional scaling. The combinations of stimuli differing in pulse rate as well as electrode number or filter frequency are shown in

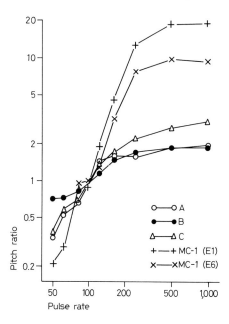

Fig. 93. Pitch ratio versus pulse rate (both on log scales) for the acoustic model and electrical stimulation. The 100 pulses/s stimulus was the standard in each case [13, 14].

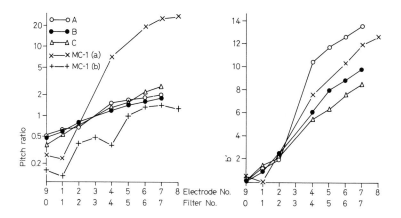

Fig. 94. Pitch ratios versus electrode or filter number and d' values versus electrode or filter number. The d' values were calculated from the pitch ratio according to the method of Braida and Durlach [18]. Subjects A, B and C tested with the acoustic model. Patients MC-1 and MC-2 using multiple-electrode cochlear implant [13, 14].

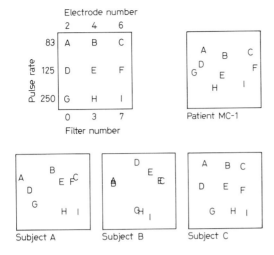

Fig. 95. Non-metric multidimensional scaling stimulus configurations with Minkowski coefficients. The top left configuration gives the key to stimulus naming. The stress value for the calculated configurations were 0.066 for MC-1, 0.103 for subject A, 0.027 for subject B, and 0.077 for subject C [13, 14].

figure 95. The spatial similarity or distance between the combinations of stimuli was measured, and showed good agreement between the acoustic model and multiple-channel electrical stimulation.

Having established that the acoustic model gave similar psychophysical results to those for multiple-channel electrical stimulation, a further study was undertaken to see if similar results could be obtained with speech perception tests. A variety of tests were administered in the hearing alone, lip-reading alone, and lip-reading plus hearing condition. The scores for the cochlear implant patient MC1, and the three normally hearing subjects using the acoustic model are shown in table 26 [13, 14]. From these results and the additional ones reported by Blamay et al. [13, 14] there was good correspondence between the speech tests for a multiple-channel cochlear implant patient using the F0-F2 speech processor and subjects using the F0-F2 model. The subjects were better lip-readers, however, and this could explain most of the differences seen, especially the CID sentence scores. The acoustic model and cochlear implant performances were also compared on the basis of the percentage information transferred for each speech feature on a 12-consonant test. The consonants were: /b, p, m, v, f, d, t, n, z, s, g, k/. The speech features were:

Table 26. Speech test scores (%) [13]

| Test | Condition | MC-1 | Subjects | | |
			A	B	C
Male-female speaker	HA	24	13	22	15
Question statement	HA	19	15	18	15
Vowels	HA	29	19	26	25
Final consonant	HA	30	27	27	34
Initial consonant	HA	21	21	33	30
AB words	HA	2	0	1	0
	LA	2	2	6	2
	HL	4	5	7	8
Phonemes	HA	12	5	6	9
	LA	12	16	23	17
	HL	21	23	25	27
CID sentences	HA	7	2	7	4
	LA	12	36	43	23
	HL	36	45	40	49
Speech tracking[1]	LA	5.5	14.5	23.5	20.0
	HL	34.0	46.6	60.7	53.8

[1] Speech tracking – words/min.
HA = Hearing or electrical stimulation alone; LA = lip-reading alone; HL = hearing or electrical stimulation plus lip-reading.

voicing, nasality, affrication, duration, and place. The results for MC1 and the average results for the three subjects are shown in table 27. From this it can be seen that the trends are very similar for both electrical stimulation and the acoustic model.

As the acoustic model on normally hearing subjects proved to be good at reproducing the speech results for the F0-F2 speech processing strategy a study using the acoustic model was undertaken to determine whether an F0-F1-F2 speech-processing strategy would give better results than the F0-F2 processor, and whether any additional information transmitted would be due to adding F1 on a place basis. An additional strategy was evaluated in which F2 was coded as rate of stimulation. The rate was proportional to the logarithm of F2 such that 1–4 kHz was mapped onto the range 50–300 pulses/s.

A confusion study on the eleven Australian vowels [15] comparing

Table 27. Percentage information transmitted [13]

Feature	MC-1			Acoustic model subjects		
	HA	LA	HL	HA	LA	HL
Voicing	36	0	41	39	2	62
Nasality	25	7	32	68	8	89
Affrication	38	43	96	31	49	98
Duration	79	30	94	85	35	97
Place	25	68	79	15	69	86
Overall	46	47	72	45	50	84

HA = Hearing or electrical stimulation alone; LA = lip-reading alone; HL = hearing or electrical stimulation and lip-reading.

Table 28. Acoustic model: Comparison of speech-processing strategies – information transmission for vowels [15]

	F2 (rate) %	F0-F2 %	F0-F1-F2 %
Total	34	56	72
Duration	83	85	95
F1 grouping	12	27	81
F2 grouping	25	68	55

the three strategies on six subjects was analysed to determine the information transmission for the vowels grouped according to duration, F1 and F2. The results of the analyses are shown in table 28. From this it can be seen that there was a definite increase in the total information transmitted for the F0-F1-F2 strategy. The F0-F1-F2 strategy was the only one that transmitted a large proportion of the F1 information. A much greater proportion of the F2 information was transmitted when coded as filter frequency rather than as pulse rate.

The information transmission for a confusion study on the consonants /p, t, k, b, d, g, m, n, s, z, v, f/ [15] was also carried out and the results are shown in table 29. The results are for the features of Miller and Nicely [103]. An additional two features were formulated to represent the main groupings of the consonant comparison matrices in a more eco-

Table 29. Acoustic model: comparison of speech processing strategies – information transmission for consonants

	Strategies		
	F2 (rate)	F0-F2	F0-F1-F2
	%	%	%
Total	37	43	49
Voicing	35	34	50
Nasality	86	84	98
Affrication	31	32	40
Duration	62	71	81
Place	19	28	28
Amplitude envelope	47	46	61
High F2	48	68	64

nomical fashion. The amplitude envelope feature classified the consonants into four groups as shown in figure 96. These groups were easily recognized by eye from the traces of the amplitude envelopes produced by the real time speech processor. The high F2 feature refers to the output of the speech processor's F2 frequency extraction circuit during the burst of the stops /t/ and /k/ or during the frication noise of /s/ and /z/. /f/ and /g/ did not give rise to the feature because the amplitude of the signal was too low during the period the F2 frequency was high. Thus the F2 feature was a binary grouping with /t, k, s, z/ in one group and the remainder of the consonants in the other. The results in table 29 show an increasing trend from the F2 (rate) to F0-F2 to F0-F1-F2 speech processors. With the F0-F1-F2 speech processor the addition of F1 information appeared to help in the transmission of all features except place while the use of F0 made very little difference to the voicing feature.

The speech processing strategies were also compared for connected discourse using the speech tracking test. The results are shown in figure 97 for three subjects. The results from the first three sessions were excluded from the analysis to minimize the effect of the initial rapid improvement of the scores. There was a significant difference between the strategies. As can be seen from figure 97 the F0-F1-F2 strategy was superior to the others, and for subjects E and F the F0-F2 strategy was superior to the F2 (rate) strategy.

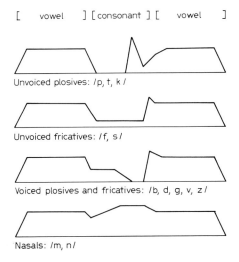

[vowel   ] [consonant] [vowel   ]

Unvoiced plosives: /p, t, k /

Unvoiced fricatives: /f, s /

Voiced plosives and fricatives: /b, d, g, v, z /

Nasals: /m, n /

Fig. 96. Schematic diagrams of the amplitude envelopes for the grouping of consonants used in the information transmission analyses. Time is the variable along the abscissa and amplitude (at the output of the AGC) along the ordinate [15].

Speech Perception Results for Speech Processor Extracting the Fundamental, First Formant, and Second Formant Frequencies

Having shown, using an acoustic model of multiple-electrode stimulation, that the F0-F1-F2 speech processing strategy was better than the F0-F2 strategy, a clinical study was then carried out to determine if the same would apply to cochlear implant patients. This would not only be a good test of the predictive value of the model, but more importantly support the rationale for the multiple-electrode speech processing strategy.

In comparing the results for the F0-F2 and F0-F1-F2 cochlear implant speech processors we considered it important to analyse the information received by the patient to see how they could use this information, and whether the information transmitted was consistent with the type of speech-processing strategy used. The percentage information transmitted for vowels and consonants for a group of 13 patients with the F0-F2 processor and seven patients with the F0-F1-F2 processor are shown in tables 30 and 31. From table 30 it can be seen that for vowels the information transmitted for duration, F1 and F2 were all greater for the F0-F1-F2 strategy. The improvement was especially large when F1 was

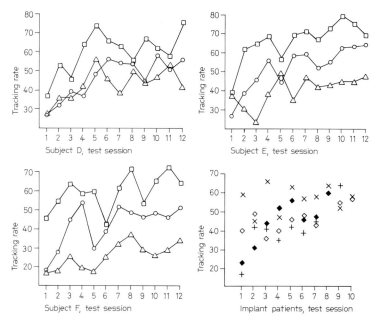

Fig. 97. Speech tracking rates (in words per minute) for the acoustic model subjects using the F2 (rate) strategy marked with △, the F0-F2 strategy marked with 0, and the F0-F1-F2 strategy, marked with □. Tracking rates for the four best patients from a clinical trial of eight cochlear implant patients using the F0-F2 strategy are shown for comparison. All results are for lip-reading plus hearing [15].

added. From table 31 it can be seen that for consonants information transmission was also better for the F0-F1-F2 speech processor compared to the F0-F2 processor for all speech features [51, 68].

If the results for information transmission for vowels and consonants for multiple-electrode stimulation (tables 30, 31) are compared with those obtained for the acoustic model (tables 28, 29) there is reasonably good agreement and the trends are similar. This therefore confirms the predictive value of the acoustic model, and supports the rationale for multiple-channel electrical stimulation.

In comparing the speech results for the F0-F2 and the F0-F1-F2 speech processors on cochlear implant patients we carried out two studies. In the first, patients who had previously used the F0-F2 speech processor were provided with the F0-F1-F2 processor, and after 2 weeks they were tested with open sets of CID sentences for electrical stimulation

Table 30. Vowels – electrical stimulation: percentage information transmitted [51, 68]

	Total	Duration	F1	F2
F0/F2 (AO) (n = 13)	51	66	31	51
F0/F1/F2 (n = 7)	64	77	51	59

Table 31. Consonants – electrical stimulation: percentage information transmitted [51, 68]

	To	Vo	NA	Af	Pl	AmEn	HiF2
F0/F2 (AO) (n = 13)	36	33	38	36	20	36	36
F0/F1/F2 (n = 7)	50	56	49	45	35	54	48

To = Total; Vo = voicing; Na = nasality; Af = affrication; Pl = place; AmEn = amplitude envelope; HiF2 = high F2.

alone. The results on seven patients using a live voice presentation of the test material are shown in figure 98. From this figure it can be seen that there was an average 100% improvement when using the F0-F1-F2 processor, and one patient achieved a 100% score for electrical stimulation alone.

When carrying out this type of comparison it is not possible to exclude a time-dependent or training effect. In other words, would the F0-F2 result have been 100% better after a further few weeks exposure? Unlikely but a remote possibility. For this reason we also carried out a comparison of the two speech-processing strategies between the two groups of patients who had similar training, ages and length of deafness. The results for open-set speech tests presented as pre-recorded material for electrical stimulation alone are shown in table 32. From this it can be seen that the improvements for the F0-F1-F2 speech processor are statistically significant for all tests. For the CID sentences there was an average 137% improvement in score and for monosyllabic words a 150% improvement. Finally, the results for tracking were also significant. There was a significant improvement in the speech tracking score which went from 28.2 words/min to 38.0 words/min for electrical stimulation combined with lip-reading minus lip-reading alone [68].

When assessing the F0-F2 and F0-F1-F2 speech processors it is also important to determine their performance in the presence of different

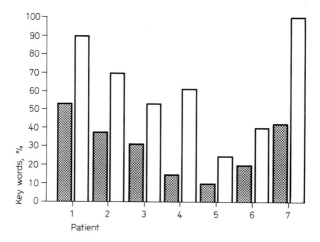

Fig. 98. Speech test results for a group of seven patients in whom the F0-F2 processor was used initially, and they were retrained to use the F0-F1-F2 processor. F0-F2 = hatch; F0-F1-F2 = clear.

levels of background noise or signal-to-noise ratios. The speech processors after all must perform in everyday conditions when there are considerable variations in background noise. The performances of the F0-F2 and F0-F1-F2 speech processor were studied using a four-choice spondee word test. The results for this study [68] show that when competing speech was used the F0-F1-F2 strategy was less affected than the F0-F2 strategy. The F0-F1-F2 strategy performed well at a signal-to-noise ratio of 10 dB. A study using the F0-F2 strategy at Iowa University [82] showed a degradation of performance at a signal-to-noise ratio of 10 dB. The results of our study have shown, however, that the F0-F1-F2 processor performs better in noise.

Speech Perception Results in Languages Other than English

Encouraging speech perception results have also been obtained in a number of patients in German [2, 94]. The 22-electrode system had been implanted in 26 patients in Hanover and has been shown to provide significant help when used in conjunction with lip-reading. It was also very encouraging that 4 months after surgery, seven of 17 patients were able to

Table 32. Speech tests: electrical stimulation alone – 3 months postsurgery

	F0/F2 (n = 13) %	F0/F1/F2 (n = 9) %
Spondees	13.6	26.0*
CID sentences	15.9	35.4*
PB words	4.9	12.4*
Phonemes	23.1	33.4*

* Significant: p = 0.01.

understand an average 42.7 words/min by speech tracking without lip-reading.

The Nucleus device has also been implanted in a Japanese patient, the results of this prosthesis on a patient [Funasaka, personal communication] have shown that the patient obtains a significant improvement for speech tracking when using the cochlear implant plus lip-reading compared to lip-reading alone. After 3.5 months training, the patient could recognize 24 bunsetsues (the minimum meaningful unit of the Japanese sentence) a minute using the cochlear implant plus lip-reading, and 14.3 bunsetsues with the lip-reading alone condition.

The Nucleus prosthesis has been implanted in a native speaker of Chinese and reported by Xu et al. [144]. In this case the patient was able to understand substantial amounts of open-set speech without lip-reading in both English and Chinese. In an open-set Chinese word recognition test he obtained 63% score for electrical stimulation alone when using words, and 86% for phonemes. The Chinese translation of open-set CID sentence tests showed an 84% score for electrical stimulation alone. Finally, when using Chinese and tested with the speech tracking procedure he obtained an 80 words/min score for electrical stimulation plus lip-reading compared to 10 words/min for lip-reading alone.

Cochlear Implants for Prelingually Deaf Adults and Children

We have defined prelingual deafness as a hearing loss occurring before the age of 4 years and postlingual deafness occurring after that age. There is now a view that there should be three categories: prelingual deafness occurring up until age 1 year, perilingual from 1 to 5 years and postlingual after 5 years.

To study the possible benefits of cochlear implantation for the prelingually deaf adults as well as children, we have been carrying out research on a small group of patients. This is being done with a multidisciplinary team. The structure of this team is as follows: audiologists, educators of the deaf, engineers, otologists, psychologists, psychophysicists, speech pathologists and speech research scientists.

Our overall management of patients is as follows: the patient is first assessed to determine whether he/she has a profound-total hearing loss. The patient is then given a trial with a hearing aid for 6 months or a 6-month course of training with a tactile aid before being reassessed. If no satisfactory progress has been made a cochlear implant is carried out. The patient is then assessed 2 months postoperatively to see what are the immediate effects of implantation. They are then reassessed at 6-month intervals.

Protocol

We have developed a protocol for the preoperative and postoperative assessment of patients, which is still in the process of evolution. Within the protocol we have laid down certain types of tests, but the details need to be varied to suit the age and language development of the patient. The protocol is shown in tables 33–35.

Table 33 shows the protocol for the assessment of speech perception. Speech perception is firstly assessed at a segmental or phonetic level for vowels and consonants. For example the PLOTT test [111] is a two-

Table 33. Speech perception – prelingual deafness assessment [53]

A Segmental – phoneme Vowels and consonants *Identification* CVC-VCV confusion studies PLOTT test [111] SERT test [17] *Feature analysis* B Suprasegmental MAC battery [109]	C Words *Closed-set* ANT [71] MST [72] NU-CHIPS [70] *Open-set* AB words [17] D Sentences BKB sentences [3] E Connected discourse Tracking [62]

Table 34. Speech production – prelingual deafness assessment [53]

A Segmental – phoneme Vowels and consonants PLE [96] CVC-VCV production and analysis B Suprasegmental PLE [96]	C Words Edinburgh articulation test [1] D Sentences Intelligibility [99] Process analysis [58, 87]

Table 35. Prelingual deafness assessment [52, 53]

1 Receptive language – semantics and syntax
PPVT [69]
Preschool language scale [145]
Development language scale [114]
2 Expressive language – semantics and syntax
  A Descriptive analysis of language samples
    Syntactic analysis
    Mean length of utterance [20]
    LARSP [60]
    Semantic analysis [16]
  B Formal language tests
    Grammatical analysis of elicited language [104–106]
3 Communication skills
  A Interactional language measures [55]

alternative first choice procedure on vowels and consonants. Supraseg-mental aspects of speech can be assessed for older children using part of the MAC battery [109]. Word perception can be assessed in young chil-dren using closed sets. For example the ANT or auditory number test [71], the MST or monosyllable spondee, trochee discrimination [72], and the NU-chips test [70] which is a 4-alternative forced choice test. Open sets of AB words [17] can be given to older children. When testing for word recognition in sentence we use the BKB sentences of Bench and Bam-ford [3].

Table 34 shows the protocol for the assessment of speech production. Speech production is assessed at a phonetic level using the phonetic level evaluation (PLE) test of Ling [96]. We also ask the patients to produce vowels in a CVC context and consonants in a VCV context. The supraseg-mental aspects of speech production are also assessed using the PLE test. Word production is assessed using the Edinburgh articulation test [1], and the ability of patients to produce words in sentences by the intelligibility test of McGarr [99] and process analysis by Ingram [87] and Crary [58].

Not only is it important to assess speech perception and production at segmental, suprasegmental and word levels, but we consider it equally important to assess receptive and expressive language and communica-tion skills. The tests for these are shown in table 35. The semantics and syntax of receptive language is assessed using the PPVT or Peabody pic-ture vocabulary test [69], the pre-school language scale of Zimmerman et al. [145] and the development language scale of Reynell [114]. The seman-tics and syntax of expressive language is assessed using the mean length of utterance, the LARSP or Language assessment remediation and screen-ing procedure of Crystal [60] and semantic analysis of Bloom and Lahey [16]. The grammatical analysis of elicited language is also carried out [104–106]. Finally communication skills are assessed using interactional language measures of Cole and St. Clair-Stokes [55].

Patient Histories

We have now operated on six prelingually deaf patients [53]. Three adults, one teenager and two children. The first patient Pre-1 was diag-nosed as having a profound-total hearing loss at age 15 months following meningitis. He had a hearing aid fitted at 15 months and was educated primarily by signing. At the time of implantation his speech was at a 25%

level on the McGarr intelligibility test and his speech reception vocabulary was at a 5-year level. He was implanted at the age of 25 years and has had the device for 1.8 years. He is an occasional user of the device.

The second patient Pre-2 was diagnosed as having a profound-total loss at the age of 3. The cause of deafness was not known but it was probably congenital. She had a hearing aid fitted at 3 and was educated primarily by signing. At the time of implantation her speech was at an 8% level on the McGarr intelligibility test and her speech reception vocabulary was at a 5-year-old level. She was implanted at the age of 24 years and has had the device for 2.5 years. She is an occasional user of the device.

The third patient Pre-3 was diagnosed as having a profound-total hearing loss at the age of 16 months due to meningitis. He was fitted with a hearing aid soon afterwards, and educated using cued speech. His speech was at a 44% level on the McGarr intelligibility test, and his speech reception vocabulary was at a 6-year-old level. He had a cochlear implant at the age of 14 years and has been wearing it for 1.5 years. He is a regular user of the device.

The fourth patient Pre-4, an adult, was diagnosed as having a severe-profound hearing loss at the age of 3 years which became a profound-total loss over the next 15 years. The cause of deafness is unknown but is presumed to be congenital. This patient is different from the first two adults as she had some residual hearing in the first 4 years of life. She is also different from the first two in that she was educated by auditory/oral means in a normal school setting and, although her speech was at the 25% level on the McGarr intelligibility test, it was more intelligible than the first two adults. She also has a much better speech reception vocabulary of 7 years 3 months. She was implanted with the device in the worse ear at the age of 22 years, and has been using it for one year. She is a regular user.

The fifth patient Pre-5 was diagnosed as having a profound-total hearing loss at the age of 3.5 years due to meningitis. He was fitted with a hearing aid soon afterwards and educated by total communication. His speech was at a 0% level on the McGarr intelligibility test and his speech reception vocabulary at a 4-year 11-month level. He had a cochlear implant at the age of 10 years and has been wearing it for one year. He is a regular user of the device and is now spending much of his time in a normal classroom setting.

The sixth patient Pre-6 was a 5-year old boy who became profoundly-totally deaf at 3 years after he developed meningitis. He commenced wearing binaural high-gain, low-frequency emphasis hearing aids soon

Table 36. Current level identification (0.4 dB difference) [138]

Subject	D′	Subject	D′
Pre-1	1.0	Pre-4	0.52
Pre-2	0.9	Post-1	3.0
Pre-3	1.7	Post-2	1.8

after recovery, and was educated by cued speech. About 6 months prior to implantation he had a trial with a 'Tactaid', and preferred to use this device to his hearing aid. Before having a cochlear implant he had a speech reception level of 2 years on the PPVT. His speech intelligibility was very poor, and he had a few one-word utterances. During the preoperative period he had some cross-modality training in scaling the intensity of light and electrotactile stimulation to help postoperatively in setting threshold and comfortable loudness and discomfort levels. The training was carried out by showing him blocks of different sizes, and training him to point to the smallest block for a low-level stimulus, and to the biggest block for an intense or uncomfortable stimulus.

Preliminary Results

Some preliminary psychophysical and speech perception results [23, 53, 138] have been obtained on these six patients. It is clear that psychophysical and speech testing is slower and more difficult on prelingually deaf patients and children than with postlingually deaf adults, as the prelingually deaf and children need more time to learn to perceive the unfamiliar sensations.

One of our first impressions with the prelingually deaf patients was that they could identify loudness changes. As this is basic to speech perception we have carried out preliminary current level identification experiments on the three adult patients and teenager. The results are shown in table 36. We have also carried out the test on two postlingually deaf patients. Their ability to identify a 0.4-dB current level difference has been measured as a d′ value. The d′ value measures the overlap between the identification of two stimuli. A high score indicates a high level of ability and vice versa for a low score.

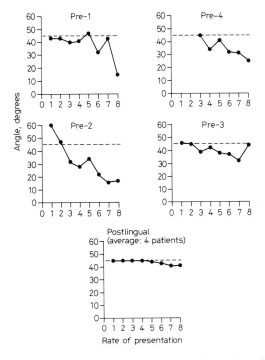

Fig. 99. A counting test. Signals counted are plotted against signals presented for rates varying from 1 to 8/s, and regression lines have been drawn to fit the data. Pre-1, Pre-2, Pre-3 and Pre-4 are prelingually deaf patients.

From table 36 it can be seen that the worst score was for Pre-4, the adult patient who had some residual hearing at birth before developing a profound-total hearing loss. The two adult prelingually deaf patients (Pre-1, Pre-2) one of whom was deafened at 15 months and both taught to sign, had scores indicating some ability, but not as good as the two postlingually deaf patients. The teenager (Pre-3) deafened at 16 months, and taught to cue, had a score close to one of the postlingually deaf patients. The results thus suggest that the prelingually deaf patients should be able to use loudness cues to obtain some speech information.

The second psychophysical test was a counting test, and this is illustrated in figure 99. Signals counted are plotted against signals presented for rates varying from 1 to 8/s, and regression lines have been drawn to fit the data. When the task is accurately performed the graph should have a

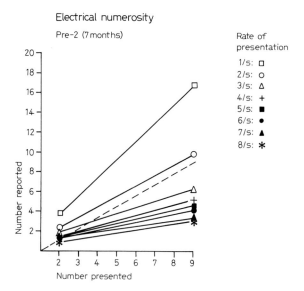

Fig. 100. A comparison of rate of presentation against the angle of the regression time for prelingually deaf patient Pre-2.

slope of 45 degrees. Notice that for higher rates of stimulation, there is greater deviation from the 45-degree slope. To make comparison easier across rates and patients we have plotted rate of presentation against the angle of the regression line. The results for the prelingually deaf patient Pre-2 are shown in figure 100. Our results showed there was only a slight fall-off in the angle of the regression line for the postlingually deaf patients. There was, however, a marked fall off with rate for the first two adult prelingually deaf patients Pre-1 and Pre-2. On the other hand, the prelingually deaf adult patient Pre-4 who had a little hearing at birth and the teenager Pre-3 had an intermediate result between the two groups.

Now the counting test is a simple overall evaluation of the auditory systems. If a patient cannot even count stimuli well it suggests there could also be something fundamentally wrong with the processing of auditory information required for speech understanding.

To further study the psychophysical abilities of the patients we carried out experiments on the identification and discrimination of different rates of stimulation. These results are shown in table 37. Here we see the identification results for the three prelingually deaf adults Pre-1, Pre-2 and Pre-4, and for the teenager Pre-3. The results for two postlingually

Table 37. Rate of stimulation discrimination [138]

Subject	Rate	Months	D'
Pre-1	75/50	3	0.6
Pre-2	75/50	3	0.3
Pre-3	150/75	1	0.3
	150/100	12	1.5
Pre-4	150/100	6	3
Post-1 and 2	75/50	>48	>3
	150/75	>48	>3

Table 38. Place of stimulation discrimination [138]

Subject	Electrode	Months	D'
Pre-1	6/14	3	0.0
Pre-2	6/14	3	0.0
Pre-3	6/14	1	0.8
		12	2.1
Pre-4	1/7	1	>3.0
	3/7	6	1.96
Post-1 and 2	6/14	>48	3.0

deaf patients are also shown. Firstly, notice that the two adult prelingually deaf patients Pre-1 and Pre-2 had much more difficulty identifying
rates of 75 and 50 pulses/s than the postlingually deaf patients. The
prelingually deaf adult Pre-4, however, performed as well as the postlingually deaf patients. The prelingually deaf teenager, Pre-3, although
given a different rate to distinguish, performed poorly at one month after
implantation. He has, however, been a regular implant user and it is
encouraging that his score has increased considerably over the 12-month
period.

These results show that a significant improvement in the perception
of stimulus rate can occur over time, and that some prelingually deaf
patients can perceive rate of stimulation as well as postlingually deaf
patients. It will be interesting to see if this ability carries over to speech
perception.

The final psychophysical results to be presented are for the identification of place of stimulation. These results are shown in table 38. The

results show that our prelingually deaf adults Pre-1 and Pre-2 had no place discrimination when tested at 3 months after implantation. This is interesting as Pre-1 had normal hearing until the age of 15 months. Our adult Pre-4 with a little hearing at birth obtained a score of 3 at one month after implantation, indicating that her ability to identify place of stimulation was similar to that of a postlingually deaf patient. At 6 months she also scored well on a task when asked to distinguish between electrodes 3 and 7. The prelingually deaf teenager, Pre-3, with hearing until 16 months and taught to cue, had a score of 0.8 at one month after implantation and this increased to 2.1 at 12 months. His results are therefore getting closer to those for the postlingually deaf patients and indicate, as with rate identification, that significant learning can take place even for a 14-year-old. It will also be interesting to see if the ability of some patients to perceive place pitch will result in better speech results.

Some preliminary speech data are also presented. In interpreting the results it is important to be aware of the fact that the patients have not had a great deal of formal training with speech perception for a variety of reasons. The first two adults (Pre-1, Pre-2) sign and are having difficulty changing their communication system. The adult with some hearing at birth (Pre-4) lives 1,000 km away, and although using the device regularly has had little formal training. The prelingually deaf teenager (Pre-3) has had formal training during the holidays, but this cannot be easily fitted into the school programme. He has visits from our staff. The prelingually deaf 10-year-old child (Pre-5) has visits from our staff. He is, however, now having most of his school lessons in a normal hearing school. His training has concentrated on an auditory-oral approach to maximize the use of the new auditory information. When placed in a total communication environment or when the implant is switched off, he will tend to revert to his previous communication mode. The prelingually deaf 5-year-old child (Pre-6) has only recently commenced his postoperative training in a cued-speech setting.

The results for a test of suprasegmental recognition (the male-female discrimination) are shown in table 39. From this it can be seen that the adult Pre-4 who had some residual hearing at birth, performed well and the teenager, Pre-3, performed better than either of the first two adults, Pre-1 and Pre-2. This can be explained on the basis of their abilities to discriminate rate of stimulation.

The results for vowel identification are shown in table 40. Here you can see the results for electrical stimulation alone, lip-reading alone and

Table 39. Supra segmental [23]

Subject	Male/female, %
Pre-1	44
Pre-2	68
Pre-3	80*
Pre-4	76*

* Significant: $p = 0.01$.

Table 40. Vowel identification (11 vowels) [23]

Subject	Months	E %	V %	EV %
PR-1	6	18	39	46
PR-2	6	14	54	50
PR-3	2	24	84	84
PR-4	1	(51)	88	93
PR-5	3	24	44	65

Parentheses include 6 vowels.

electrical stimulation plus lip-reading. Notice firstly all patients are able to obtain some vowel identification for electrical stimulation. For patients Pre-1 and Pre-2 it is just above chance and most likely due to amplitude cues. The scores for the other patients are better and this suggests that they may be using place pitch information. This is also consistent with the psychophysical scores. However, for patient Pre-4 there is little improvement when electrical stimulation is combined with lip-reading compared to lip-reading alone. The lip-reading score is high in Pre-3 and 4 so further improvements are not likely. The increased scores for lip-reading combined with electrical stimulation compared to lip-reading alone in our children Pre-5 are quite encouraging.

The results for consonant identification are shown in table 41. Here you can see that the scores for Pre-4 and Pre-5 are above chance for electrical stimulation alone. There is also some improvement when electrical stimulation is combined with lip-reading compared to lip-reading alone.

Table 41. Consonant identification (12 consonants) [23]

Subjects	Months	E %	V %	EV %
PR-1	6	12	33	40
PR-2	6	8	26	30
PR-3	1	7	31	31
PR-4	2	(31)	47	65
PR-5	3	17	42	50

Parentheses include 6 consonants.

Table 42. Pre-4 (6 months) [23]

Tests	%
CNC words	
Lip-reading alone	30
Electrical + lip-reading	56*
CID sentences	
Lip-reading alone	43
Electrical + lip-reading	80*

* $p > 0.01$.

The improvement in the combined condition for Pre-1 and Pre-2 may be due to their ability to use amplitude information.

Finally, how do all these tests relate to the patients' perception of words and running speech? These tests are underway at the moment. Firstly, for adults Pre-1 and Pre-2 their language level and communication skills were so poor to start with their perception of words and running speech is still very difficult to assess. The teenager Pre-3 has recently been undertaking tracking tests and the results for his first five sessions have shown an average 30% improvement in tracking rate from 25.4 to 33 words/min when using the cochlear implant combined with lip-reading compared with lip-reading alone. This is an encouraging result as it indicates that a 14-year-old can obtain benefit in the perception of running speech. It is also interesting that his place and rate discrimination improved over 6 months and could account for this result. The adult

Pre-4 who had some hearing at birth before developing a profound-total hearing loss had the CNC word and CID sentences tests administered with lip-reading alone and lip-reading plus electrical stimulation. The results at 6 months postoperatively are shown in table 42. The results were significant at a 0.01 probability level.

Finally, although children Pre-5 and Pre-6 are doing well at school, word and running speech tests have not yet been done because they have only fairly recently had an implant, and more language training is required before they can be adequately assessed.

Summary

To summarize, our preliminary results indicate that some prelingually deaf patients may get worthwhile help from a multiple-electrode cochlear implant which extracts formants. They can understand words and running speech better when using the cochlear implant with lip-reading compared to lip-reading alone. It has been encouraging that these improvements can occur in young adults and teenagers. It has also been encouraging that some can recognize place pitch as well as rate pitch. There are, however, considerable variations in performance and this may be due to the following factors: whether they have had some hearing after birth, the method of education used, the motivation of the patient and age at implantation.

In conclusion it is important to emphasize that deaf children are severely disadvantaged however good their teacher of the deaf. Research on cochlear implants offers hope for profoundly-totally deaf children. These developments will not replace the caring, competent educators but complement their skills. There is also a greater need than ever for an interdisciplinary approach to these children.

References

1 Anthony, A.; Bogle, D.; Ingram, T.T.S.; McIsaac, M.W.: The Edinburgh articulation test (Churchill-Livingstone, Edinburgh 1971).

2 Battmer, R.D.; Lehnhardt, E.; Laszig, R.: Hannover cochlear implant program using the Nucleus device. Int. Cochlear Implant Symp. and Workshop, Melbourne 1985. Ann. Otol. Rhinol. Lar., suppl. *128, vol. 96:* 129–132 (1987).

3 Bench, J.; Bamford, J.: Speech-hearing tests and the spoken language of hearing-impaired children (Academic Press, London 1979).

4 Berkowitz, R.G.; Franz, B.K.-H.; Shepherd, R.K.; Clark, G.M.; Bloom, D.M.: Pneumococcal middle ear infection and cochlear implantation. Int. Cochlear Implant Symp. and Workshop, Melbourne 1985. Ann. Otol. Rhinol. Lar., suppl. *128, vol. 96:* 55–57 (1987).

5 Berkowitz, R.G.; Franz, B.K.-H.; Shepherd, R.K.; Clark, G.M.; Bloom, D.M.: Cochlear implant and otitis media: a pilot study to assess the feasibility of *Pseudomonas aeroginosa* and *Streptococcus pneumoniae* infection in the cat. J. Otolaryngol. Soc. Aust. *5:* 297–299 (1985).

6 Bernard, J.R.L.: Toward the acoustic specification of Australian English. Z. Phonet. *23:* 113–128 (1970).

7 Black, R.C.; Clark, G.M.: Electrical network properties and distribution of potentials in the cat cochlea. Proc. Aust. Physiol. Pharmacol. Soc. *9:* 71 (1978).

8 Black, R.C.; Hannaker, P.: Dissolution of smooth platinum electrodes in biological fluids. Appl. Neurophysiol. *42:* 366–374 (1979).

9 Black, R.C.; Clark, G.M.: Differential electrical excitation of the auditory nerve. J. acoust. Soc. Am. *67:* 868–874 (1980).

10 Black, R.C.; Clark, G.M.; Patrick, J.F.: Current distribution measurements within the human cochlea. IEEE Trans. biomed. Engng *28:* 721–725 (1981).

11 Black, R.C.; Steel, A.C.; Clark, G.M.: Amplitude and pulse rate difference limens for electrical stimulation of the cochlea following graded degeneration of the auditory nerve. Acta oto-lar. *95:* 27–33 (1983).

12 Black, R.C.; Clark, G.M.; O'Leary, S.J.; Walters, C.: Intracochlear electrical stimulation of normal and deaf cats investigated using brainstem response audiometry. Acta oto-lar. suppl. 399, pp. 5–17 (1983).

13 Blamey, P.J.; Dowell, R.C.; Tong, Y.C.; Brown, A.M.; Luscombe, S.M.; Clark, G.M.: Speech processing studies using an acoustic model of a multiple-channel cochlear implant. J. acoust. Soc. Am. *76:* 104–110 (1984).

14 Blamey, P.J.; Dowell, R.C.; Tong, Y.C.; Clark, G.M.: An acoustic model of a multiple-channel cochlear implant. J. acoust. Soc. Am. *76:* 97–103 (1984).

15 Blamey, P.J.; Martin, L.F.A.; Clark, G.M.: A confusion of three speech coding strategies using an acoustic model of a cochlear implant. J. acoust. Soc. Am. *77:* 209–217 (1985).

16   Bloom, L.; Lahey, M.: Language development and language disorders (Wiley&Sons, Toronto 1978).

17   Boothroyd, A.: Developments in speech audiometry. Sound *2:* 3 (1968).

18   Braida, L.D.; Durlach, N.I.: Intensity perception. II. Resolution in our internal paradigms. J. acoust. Soc. Am. *51:* 483–502 (1972).

19   Brennan, W.J.; Clark, G.M.: An animal model of acute otitis media and the histopathological assessment of a cochlear implant in the cat. J. Lar. Otol. *99:* 851–856 (1985).

20   Brown, R.: A first language (Allen&Unwin, London 1973).

21   Brown, A.M.; Clark, G.M.; Dowell, R.C.; Martin, L.F.A.; Seligman, P.M.: Telephone use by a multi-channel cochlear implant. An evaluation using open-set CID sentences. J. Lar. Otol. *99:* 231–238 (1985).

22   Bruck, S.D.: Properties of biomaterials in the physiological environment (CRC Press, Cleveland 1980).

23   Busby, P.A.; Tong, Y.C.; Clark, G.M.: Speech perception studies in the first year of usage of a multiple-electrode cochlear implant by prelingual patients. J. acoust. Soc. Am. *80:* suppl. 1, p. 530 (1986).

24   Brummer, S.B.; McHardy, J.; Turner, M.J.: Electrical stimulation with Pt electrodes. Trace analysis for dissolved platinum and other dissolved electrochemical products. Brain Behav. Evol. *14:* 10–22 (1977).

25   Brummer, S.B.; Turner, M.J.: Electrical stimulation with Pt electrodes. I. A method for determination of 'real' electrode areas. IEEE Trans. biomed. Engng *24:* 436 (1977).

26   Clark, G.M.: Responses of cells in the superior olivary complex of the cat to electrical stimulation of the auditory nerve. Expl Neurol. *24:* 124–136 (1969).

27   Clark, G.M.: A neurophysiological assessment of the surgical treatment of perceptive deafness. Int. Audiol. *9:* 103–109 (1970).

28   Clark, G.M.; Dunlop, C.W.: A field potential study of inhibition in the cat superior olivary complex. J. audit. Res. *11:* 79–87 (1971).

29   Clark, G.M.; Nathar, J.M.; Kranz, H.G.; Maritz, J.S.: A behavioural study on electrical stimulation of the cochlea and central auditory pathways of the cat. Expl Neurol. *36:* 350–361 (1972).

30   Clark, G.M.; Kranz, H.G.; Minas, H.J.: Behavioral thresholds in the cat to frequency modulated sound and electrical stimulation of the auditory nerve. Expl Neurol. *41:* 190–200 (1973).

31   Clark, G.M.: A hearing prosthesis for severe perceptive deafness – experimental studies. J. Lar. Otol. *87:* 929–945 (1973).

32   Clark, G.M.; Kranz, H.G.; Nathar, J.M.: Histopathological findings in cochlear implants in cats. J. Lar. Otol. *89:* 945–504 (1975).

33   Clark, G.M.; Black, R.; Dewhurst, D.J.; Forster, I.C.; Patrick, J.F.; Tong, Y.C.: A multiple-electrode hearing prosthesis for cochlear implantation in deaf patients. Med. Prog. Technol. *5:* 127–140 (1977).

34   Clark, G.M.: An evaluation of per-scalar cochlear electrode implantation techniques: a histopathological study. J. Lar. Otol. *91:* 185–199 (1977).

35   Clark, G.M.; Tong, Y.C.; Black, R.; Forster, I.C.; Patrick, J.F.; Dewhurst, D.J.: A multiple electrode cochlear implant. J. Lar. Otol. *91:* 935–945 (1977).

36   Clark, G.M.; Tong, Y.C.; Bailey, Q.R.; Black, R.C.; Martin, L.F.; Millar, J.B.; O'Loughlin, B.J.; Patrick, J.F.; Pyman, B.C.: A multiple-electrode cochlear implant. J. Otolaryngol. Soc. Aust. *4:* 208–212 (1978).

37   Clark, G.M.; Pyman, B.C.; Bailey, Q.E.: The surgery for multiple-electrode cochlear implantation. J. Lar. Otol. *93:* 215–223 (1979).

38   Clark, G.M.; Pyman, B.C.; Pavillard, R.E.: A protocol for the prevention of infection in cochlear implant surgery. J. Lar. Otol. *94:* 1377–1386 (1980).

39   Clark, G.M.; Tong, Y.C.; Martin, L.F.A.: A multiple-channel cochlear implant: an evaluation using closed-set spondaic words. J. Lar. Otol. *95:* 461–464 (1981).

40   Clark, G.M.; Tong, Y.C.; Martin, L.F.A.; Busby, P.A.: A multiple-channel cochlear implant: an evaluation using an open-set word test. Acta oto-lar. *91:* 173–175 (1981).

41   Clark, G.M.; Tong, Y.C.: A multiple-channel cochlear implant – a summary of results for two patients. Archs Otolar. *108:* 214–217 (1982).

42   Clark, G.M.; Dowell, R.C.; Brown, A.M.; Luscombe, S.M.; Pyman, B.P.; Webb, R.L.; Bailey, Q.R.; Seligman, P.M.: The clinical trial of a multi-channel cochlear prosthesis. An initial study on four patients with a profound-to-total hearing loss. Med. J. Aust. *2:* 430–433 (1983).

43   Clark, G.M.; Crosby, P.A.; Dowell, R.C.; Kuzma, J.A.; Money, D.M.; Patrick, J.F.; Seligman, P.M.; Tong, Y.C.: The preliminary clinical trial of a multi-channel cochlear implant hearing prosthesis J. acoust. Soc. Am. *74:* 1911–1913 (1983).

44   Clark, G.M.; Shepherd, R.K.; Patrick, J.F.; Black, R.C.; Tong, Y.C.: Design and fabrication of the banded electrode array. Cochlear prosthesis; in Parkins, Anderson, International Symposium. Ann. N.Y. Acad. Sci. *405:* 191–201 (1983).

45   Clark, G.M.; Tong, Y.C.; Dowell, R.C.: Comparison of two cochlear implant speech-processing strategies. Ann. Otol. Rhinol. Lar. *93:* 127–131 (1984).

46   Clark, G.M.; Dowell, R.C.; Pyman, B.C.; Brown, A.M.; Webb, R.L.; Tong, Y.C.; Bailey, Q.; Seligman, P.M.: Clinical trial of a multi-channel cochlear prosthesis: results on 10 postlingually deaf patients. Aust. N.Z. J. Surg. *54:* 519–526 (1984).

47   Clark, G.M.; Tong, Y.C.; Patrick, J.F.; Seligman, P.M.; Crosby, P.A.; Kuzma, J.A.; Money, D.K.: A multi-channel cochlear prosthesis for profound-to-total hearing loss. J. med. Eng. Technol. *8:* 3–8 (1984).

48   Clark, G.M.; Pyman, B.C.; Webb, R.L.; Bailey, Q.E.; Shepherd, R.K.: Surgery for an improved multiple-channel cochlear implant. Ann. Otol. Rhinol. Lar. *93:* 204–207 (1984).

49   Clark, G.M.; Shepherd, R.K.: Cochlear implant round window sealing procedure in the cat. Acta oto-lar., suppl. 410, pp. 5–15 (1984).

50   Clark, G.M.; Tong, Y.C.: The engineering of future cochlear implants; in Gray, Cochlear implants (Croon Helm, London 1985).

51   Clark, G.M.: The University of Melbourne/Cochlear Corporation (Nucleus) Program; in Balkany, The cochlear implant. Otolarnygol. Clin. North Am. *19* (1986).

52   Clark, G.M.; Pyman, B.C.; Webb, R.L.; Franz, B.; Redhead, T.J.; Shepherd, R.K.: The surgery for the insertion and reinsertion of the banded electrode array. Int. Cochlear Implant Symp. and Workshop, Melbourne 1985. Ann. Otol. Rhinol. Lar., suppl. *128, vol. 96:* 10–12 (1987).

53   Clark, G.M.; Busby, P.A.; Roberts, S.A.; Dowell, R.C.; Tong, Y.C.; Blamey, P.J.; Nien-

huys, T.G.; Mecklenburg, D.J.; Webb, R.L.; Pyman, B.C.; Franz, B.K.: Preliminary results for the Cochlear Corporation multi-electrode intracochlear implant on six prelingually deaf patients. Am. J. Otol. *8:* 234–239 (1987).

54 Clifford, A.; Gibson, W.: The anatomy of the round window with respect to cochlear implant insertion. Int. Cochlear Implant Symp. and Workshop, Melbourne 1985. Ann. Otol. Rhinol. Lar., suppl. *128, vol. 96:* 17–19 (1987).

55 Cole, E.B.; St. Clair-Stokes, J.: Caregiver-child interactive behaviours: a clinical procedure for the development of spoken language in hearing-impaired children. Br. J. Audiol. *18:* 7–16 (1984).

56 Cooper, F.S.; Delattre, P.C.; Liberman, A.M.; Borst, J.M.; Gerstman, L.J.: Some experiments on the perception of synthetic speech sounds. J. acoust. Soc. Am. *24:* 597–606 (1952).

57 Cranswick, N.E.; Franz, B.K.-H.; Clark, G.M.; Shepherd, R.K.; Bloom, D.M.: Middle ear infection postimplantation: the response of the round window membrane to *Streptococcus pyogenes.* Int. Cochlear Implant Symp. and Workshop, Melbourne 1985. Ann. Otol. Rhinol. Lar., suppl. *128, vol. 96:* 53–54 (1987).

58 Crary, M.A.: Phonological interaction concepts and procedures (College Hill Press, San Diego 1981).

59 Crosby, P.A.; Seligman, P.M.; Patrick, J.F.; Kuzma, J.A.; Money, D.K.; Ridler, J.; Dowell, R.: The Nucleus multi-channel implantable hearing prosthesis. Acta oto-lar., suppl. 411, pp. 111-114 (1984).

60 Crystal, D.: Working with LARSP (Arnold, London 1979).

61 Davis, H.; Silverman, S.R.: Hearing and deafness; 3rd ed., p. 216 (Holt, Rinehard&Winston, New York 1970).

62 De Filippo, C.L.; Scott, B.L.: A method for tracking and evaluating the reception of ongoing speech. J. acoust. Soc. Am. *63:* 1186–1192 (1978).

63 Dowell, R.C.; Martin, L.F.A.; Tong, Y.C.; Clark, G.M.; Seligman, P.M.; Patrick, J.F.: A 12-consonant confusion study on a multiple-channel cochlear implant patient. J. Speech Hear. Res. *25:* 209–516 (1982).

64 Dowell, R.C.; Webb, R.L.; Clark, G.M.: Clinical results using a multiple-channel cochlear prosthesis. Acta oto-lar., suppl. 411, pp. 230–236 (1984).

65 Dowell, R.C.; Martin, L.F.A.; Clark, G.M.; Brown, A.M.: Results of a preliminary clinical trial on a multiple channel cochlear prosthesis. Ann. Otol. Rhinol. Lar., *94:* 244–250 (1985).

66 Dowell, R.C.; Tong, Y.C.; Blamey, P.J.; Clark, G.M.: Psychophysics of multiple-channel stimulation, in Schindler, Merzenich, Cochlear implants, pp. 283–290 (Rowen Press, New York 1985).

67 Dowell, R.C.; Mecklenburg, D.J.; Clark, G.M.: Speech recognition for 40 patients receiving multi-channel cochlear implants. Archs. Otolar. *112:* 1054–1059 (1986).

68 Dowell, R.C.; Seligman, P.M.; Blamey, P.J.; Clark, G.M.: Speech perception using a two formant 22-electrode cochlear prosthesis in quiet and in noise. Acta oto-lar. (in press).

69 Dunn, L.M.; Dunn, L.M.: Peabody picture vocabulary test (Am. Guidance Services, Circle Pines 1981).

70 Elliott, L.L.; Katz, D.R.: Development of a new children's test of speech discrimination (Auditec, St. Louis 1981).

71 Erber, N.P.: Use of auditory numbers tests to evaluate speech perception abilities of hearing-impaired children. J. Speech Hear. Dis. *45:* 527–532 (1980).

72 Erber, N.P.; Alancewicz, C.M.: Audiologic evaluation of deaf children. J. Speech Hear. Dis. *41:* 256–267 (1976).

73 Evans, E.F.: The frequency response and other properties of single fibres in the guinea-pig cochlear nerve. J. Physiol., Lond. *226:* 263–287 (1972).

74 Fant, C.G.M.: On the predictability of formant levels and spectrum envelopes from formant frequencies; in Halle, Lunt, MacLean, For Roman Jakobson (Mouton, The Hague 1956).

75 Forster, I.C.: Theoretical design and implementation of a transcutaneous, multichannel stimulator for neural prosthesis applications. J. biomed. Engng *3:* 107–120 (1981).

76 Franz, B.; Clark, G.M.; Bloom, D.M.: Permeability of the implanted round window membrane. Acta oto-lar., suppl. 410, pp. 17–23 (1984).

77 Franz, B.K.-H.; Clark, G.M.; Bloom, D.M.: Cochlear implants and otitis media: a comparison of cochlear implant electrode insertion techniques, and the effects of otitis media induced with group A streptococci. Ann. Otol. Rhinol. Lar. *96:* 174–177 (1987).

78 Franz, B.K.; Clark, G.M.: Refined surgical technique for insertion of banded electrode array. Int. Cochlear Implant Symp. and Workshop, Melbourne 1985. Ann. Otol. Rhinol. Lar., suppl. *128, vol. 96:* 15–17 (1987).

79 Franz, B.K.-H.; Clark, G.M.; Bloom, D.M.: Surgical anatomy of the round window with special reference to cochlear implantation. J. Lar. Otol. *101:* 97–102 (1987).

80 Frazier, L.F.; Elliott, D.M.: Size and reliability of frequency-difference thresholds determined with operant tracking procedure. J. exp. Anal. Behav. *6:* 189–192 (1963).

81 Galambos, R.; Davis, H.: Responses of single auditory nerve fibers to acoustic stimulation. J. Neurophysiol. *6:* 39–57 (1943).

82 Gantz, B.J.; McCabe, B.F.; Tyler, R.J.; Preece, J.P.: Evaluation of four cochlear implant designs. Int. Cochlear Implant Symp. and Workshop, Melbourne 1985. Ann. Otol. Rhinol. Lar., suppl. *128, vol. 96:* 145–147 (1987).

83 Giolas, T.G.; Cooker, N.S.; Duffy, J.R.: The predictability of words in sentences. J. audit. Res. *10:* 328–334 (1970).

84 Halsey, R.J.; Swaffield, J.: Analysis-synthesis telephony, with special reference to the vocoder. Inst. Elec. Engrs, Lond. *95:* 391–411 (1948).

85 Hambrecht, F.T.: Neural prostheses. Annu. Rev. Biophys. Bioeng. *8:* 239–267 (1979).

86 House, W.F.; Edgerton, B.J.: A multiple-electrode cochlear implant. Ann. Otol. Rhinol. Lar., suppl. 91, pp. 104–116 (1982).

87 Ingram, I.: Procedures for the phonological analysis of children's language (University Park Press, Baltimore 1981).

88 Javel, E.; Tong, Y.C.; Shepherd, R.K.; Clark, G.M.: Responses of cat auditory nerve fibres to biphasic electrical current pulses. Int. Cochlear Implant Symp. and Workshop, Melbourne 1985. Ann. Otol. Rhinol. Lar., suppl. *128, vol. 96:* 26–30 (1987).

89 Johnsson, L.G.; House, W.F.; Linthicum, I.H.: Otopathological findings in a patient with bilateral cochlear implants. Ann. Otol. Rhinol. Lar., suppl. 91, pp. 74–90 (1982).

90 Kiang, N.Y.-S.: Discharge patterns of single fibers in the cat's auditory nerve. MIT Research Monogr. 35 (MIT Press, Cambridge 1965).

91 Kiang, N.Y.S.: Processing of speech by the auditory nervous system. J. acoust. Soc. Am. *68:* 830–835 (1980).

92 Koenig, W.; Dunn, H.K.; Lacy, L.Y.: The sound spectrograph. J. acoust. Soc. Am. *17:* 19–49 (1946).

93 Laird, R.K.: The bioengineering development of a sound encoder for an implantable hearing prosthesis for the profoundly deaf; master of engineering science thesis, Melbourne (1979).

94 Lehnhardt, E.; Battmer, R.D.; Nakahodo, K.; Laszig, R.: Cochlear implants. HNO *34:* 271–279 (1986).

95 Liboff, A.R.; Rinaldi, R.A.; Lavine, L.S.; Shamos, M.H.: On electrical conduction in living bone. Clin. Orthop. *106:* 330–335 (1975).

96 Ling, D.: Speech and the hearing impaired child: theory and practice (Alexander Graham Bell Association for the Deaf, Washington 1976).

97 Martin, L.F.A.; Tong, Y.C.; Clark, G.M.: A multiple-channel cochlear implant. Archs Otolar. *107:* 157–159 (1981).

98 Martin, L.F.A.: Evaluation of speech processing strategies for patients with a implanted hearing prosthesis; MSc. thesis, Melbourne (1985).

99 McGarr, N.: The intelligibility of deaf speech to experienced and inexperienced listeners. J. Speech Hear. Res. *26:* 451–458 (1983).

100 Merzenich, M.M.: Studies on electrical stimulation of the auditory nerve in animals and man: cochlear implants; in Tower, The nervous system, vol. 3, pp. 337–548 (Raven Press, New York 1974).

101 Merzenich, M.M.; White, M.: Coding considerations in design of cochlear prostheses. Ann. Otol. Rhinol. Lar. *89:* 84–87 (1980).

102 Millar, J.B.; Tong, Y.C.; Clark, G.M.: Speech processing for cochlear implant prostheses. J. Speech Hear. Res. *27:* 280–296 (1984).

103 Miller, G.A.; Nicely, P.E.: An analysis of perceptual confusion among some English consonants. J. acoust. Soc. Am. *27:* 338–352 (1955).

104 Moog, J.S.; Geers, A.E.: Grammatical analysis of elicited language – simple sentence level (Central Institute for the Deaf, St. Louis 1979).

105 Moog, J.S.; Geers, A.E.: Grammatical analysis of elicited language – complex sentence level (Central Institute for the Deaf, St. Louis 1980).

106 Moog, J.S.; Geers, A.E.: Grammatical analysis of elicited language – pre-sentence level (Central Institute for the Deaf, St. Louis 1983).

107 Mullison, E.G.: Current studies of silicones in plastic surgery. Archs Otolar. *83:* 59 (1966).

108 O'Leary, S.J.; Black, R.C.; Clark, G.M.: Current distributions in the cat cochlea: a modelling and electrophysiolgical study. Hear. Res. *18:* 273–281 (1985).

109 Owens, E.; Kessler, D.K.; Schubert, E.D.: The minimal auditory capabilities (MAC) battery. Hear. Aid J. *34:* 9–34 (1981).

110 Patrick, J.F.; MacFarlane, J.C.: Comparative mechanical properties of single and multi-channel electrodes. Int. Cochlear Implant Symp. and Workshop, Melbourne 1985. Ann. Otol. Rhinol. Lar., suppl. *128, vol. 96:* 46–48 (1987).

111 Plant, G.: A diagnostic speech test for severely and profoundly hearing impaired children. Aust. J. Audiol. *6:* 1–9 (1984).

112 Pyman, B.C.; Clark, G.M.; Dowell, R.C.; Webb, R.L.; Brown, A.M.; Bailey, Q.E.; Luscombe, S.M.: The clinical trial of a multi-channel cochlear prosthesis. J. Otolaryngol. Soc. Aust. *5:* 43–46 (1983).

113 Reddy, G.N.; Saha, S.: Electrical and dielectric properties of wet bone as a function of frequency. IEEE Trans. biomed. Engng *31:* 296–303 (1984).

114 Reynell, J.K.: Reynell developmental language scales manual (NFER-Nelson, Windsor 1983).

115 Rose, J.E.; Galambos, R.; Hughes, J.R.: Microelectrode studies of the cochlear nuclei of the cat. Johns Hopkins Hosp. Bull. *104:* 211–251 (1959).

116 Rose, J.E.; Greenwood, D.D.; Goldberg, J.M.; Hind, J.E.: Some discharge characteristics of single neurons in the inferior colliculus of the cat. Tonotopical organization, relation of spike counts to tone intensity and firing patterns of single elements. J. Neurophysiol. *26:* 294–320 (1963).

117 Rose, J.E.; Brugge, J.F.; Anderson, D.J.; Hind, J.E.: Phase-locked response to low-frequency tones in single auditory nerve fibers of the squirrel monkey. J. Neurophysiol. *30:* 767–793 (1967).

118 Sachs, M.B.; Young, E.D.: Effects of nonlinearities on speech encoding in the auditory nerve. J. acoust. Soc. Am. *68:* 858–875 (1980).

119 Schindler, R.A.; Merzenich, M.M.; White, M.W.; Bjorkroth, B.: Multi-electrode intracochlear implants: neural survival and stimulation patterns. Archs Otolar. *103:* 691–699 (1977).

120 Shepherd, R.K.; Clark, G.M.; Black, R.C.: Chronic electrical stimulation of the auditory nerve in cats. Acta oto-lar. suppl. 399, pp. 19–31 (1983).

121 Shepherd, R.K.; Webb, R.L.; Clark, G.M.; Pyman, B.C.; Hirshorn, M.S.; Murray, M.T.; Houghton, M.E.: Implanted material tolerance studies for a multiple-channel cochlear prosthesis. Acta oto-lar., suppl. 411, pp. 71–81 (1984).

122 Shepherd, R.K.; Clark, G.M.; Pyman, B.C.; Webb, R.L.: The banded intracochlear electrode array: an evaluation of insertion trauma. Ann. Otol. Rhinol. Lar. *94:* 55–59 (1985).

123 Shepherd, R.K.; Murray, M.T.; Houghton, M.E.; Clark, G.M.: Scanning electron microscopy of chronically stimulated platinum intracochlear electrodes. Biomaterials *6:* 237–242 (1985).

124 Shower, E.G.; Biddulp, R.: Differential pitch sensitivity of the ear. J. acoust. Soc. Am. *2:* 275–287 (1931).

125 Simmons, F.B.: Electrical stimulation of the auditory nerve in man. Archs Otolar. *84:* 24–76 (1966).

126 Simmons, F.B.: Permanent intracochlear electrodes in cats, tissue tolerance and cochlear microphonics. Laryngoscope, St. Louis *77:* 171–186 (1967).

127 Stevens, K.N.; House, A.S.: Development of a quantitative description of vowel articulation. J. acoust. Soc. Am. *27:* 484–493 (1955).

128 Sutton, D.; Millar, J.M.; Pfingst, B.E.: Comparison of cochlear histopathology following two implant designs for use in scala tympani. Ann. Otol. Rhinol. Lar., suppl. 89, pp. 11–14 (1980).

129 Thornton, A.R.; Raffin, M.J.M.: Speech-discrimination scores modelled as a binomial variable. J. Speech Hear. Res. *21:* 507–518 (1978).

130 Tong, Y.C.; Black, R.C.; Clark, G.M.; Forster, I.C.; Miller, J.B.; O'Loughlin, B.J.; Patrick, J.F.: A preliminary report on a multiple-channel cochlear implant operation. J. Lar. Otol. *93:* 679–695 (1979).

131 Tong, Y.C.; Clark, G.M.; Seligman, P.M.; Patrick, J.F.: Speech processing for a

multiple-electrode cochlear implant hearing prosthesis. J. acoust. Soc. Am. *68:* 1897–1899 (1980).

132 Tong, Y.C.; Miller, J.B.; Clark, G.M.; Martin, L.F.A.; Busby, P.A.: Psychophysical studies for a multiple-channel cochlear implant. Proc. 10th Int. Congr. on Acoustics, B-12.4 (1980).

133 Tong, Y.C.; Clark, G.M.; Dowell, R.C.; Martin, L.F.A.; Seligman, P.M.; Patrick, J.F.: A multiple-channel cochlear implant and wearable speech-processor. Acta oto-lar. *92:* 193–198 (1981).

134 Tong, Y.C.; Clark, G.M.; Blamey, P.J.; Busby, P.A.; Dowell, R.C.: Psychophysical studies for two multiple-channel cochlear implant patients. J. acoust. Soc. Am. *71:* 153–160 (1982).

135 Tong, Y.C.; Blamey, P.J.; Dowell, R.C.; Clark, G.M.: Psychophysical studies evaluating the feasibility of a speech processing strategy for a multiple-channel cochlear implant. J. acoust. Soc. Am. *74:* 73–80 (1983).

136 Tong, Y.C.; Dowell, R.C.; Blamey, P.J.; Clark, G.M.: Two-component hearing sensations produced by two-electrode stimulation in the cochlea of a totally deaf patient. Science *219:* 993–994 (1983).

137 Tong, Y.C.; Clark, G.M.: Absolute identification of electric pulse rates and electrode positions by cochlear implant patients. J. acoust. Soc. Am. *77:* 1881–1888 (1985).

138 Tong, Y.C.; Busby, P.A.; Clark, G.M.: Psychophysical studies on prelingual patients using a multiple-electrode cochlear implant. J. acoust. Soc. Am. *80:* suppl. 1, p. 530 (1986).

139 Tsuchitani, G.; Boudreau, J.C.: Encoding of stimulus frequency and intensity by cat superior olive S-segment cells. J. acoust. Soc. Am. *42:* 794–805 (1967).

140 Wever, E.G.; Bray, C.W.: Auditory nerve impulses. Science *71:* 215 (1930).

141 Williams, A.J.; Clark, G.M.; Stanley, G.V.: Pitch discrimination in the cat through electrical stimulation of the terminal auditory nerve fibers. Physiol. Psychol. *4:* 23–27 (1976).

142 Williams, D.F.; Roaf, R.: Implants in surgery (Saunders, London 1973).

143 Wilson, J.P.: An auditory after image; in Plomp, Smorgenburg, Frequency analysis and periodicity detection in hearing, p. 303 (Sijthoff, Leiden 1970).

144 Xu, S.A.; Dowell, R.C.; Clark, G.M.: Results for Chinese and English in a multi-channel cochlear implant patient. Int. Cochlear Implant Symp. and Workshop, Melbourne 1985. Ann. Otol. Rhinol. Lar., suppl. 128, *vol. 96:* 126–127 (1987).

145 Zimmerman, I.L.; Steiner, V.G.; Pond, R.E.: Preschool language scales manual (Merrill, Columbus 1979).

146 FDA: Electromagnetic compatibility standard for medical revues, MDS-201-0004. Food and Drug Administration, National Technical Information Service, Accession No. PB80-180284 (FDA, Washington 1979).

147 AS 1099 – Basic environmental testing procedures for electronics and telecommunications purposes. Australian Standard.

148 The United States Pharmacopeia (1970).

149 The United States Pharmacopeia; 20th revision, United States Pharmacopeial Convention, Inc., Rockville, Maryland, pp. 950–953 (1980).

Subject Index